Design Knowing and Learning: Cognition in Design Education

ELSEVIER SCIENCE INTERNET HOMEPAGE:

http://www.elsevier.nl
http://www.elsevier.com
http://www.elsevier.jp

Consult homepage for full catalogue information on all books, journals and electronic products and services

Design Knowing and Learning: Cognition in Design Education

Edited by

Charles M. Eastman
W. Michael McCracken
Wendy C. Newstetter

Georgia Institute of Technology
Atlanta GA, USA

2001

ELSEVIER

Amsterdam – London – New York – Oxford – Paris – Shannon – Tokyo

ELSEVIER SCIENCE Ltd
The Boulevard, Langford Lane
Kidlington, Oxford OX5 1GB, UK

First edition 2001

Library of Congress Cataloging in Publication Data
A catalog record from the Library of Congress has been applied for.

British Library Cataloguing in Publication Data
A catalogue record from the British Library has been applied for.

ISBN: 0 08 043868 7

⊗ The paper used in this publication meets the requirements of ANSI/NISO Z39.48-1992 (Permanence of Paper).
Printed in The Netherlands.

Contributors

Omer Akin
School of Architecture
Carnegie Mellon University
Pittsburgh, PA 15213
Email: oa04@andrew.cmu.edu

Cynthia J. Atman
Center for Engineering Learning and Teaching
Industrial Engineering
University of Washington
Seattle, WA 98195
Email: atman@engr.washington.edu

Louis L. Bucciarelli
School of Engineering
Massachusetts Institute of Technology
Cambridge, MA 02139
Email: llbjr@mit.edu

John M. Carroll
Center for Human-Computer Interaction
Computer Science Department
Virginia Polytechnic Institute and State University
636 McBryde Hall, , Blacksburg, VA 24061-0106
Email: carroll@cs.vt.edu

David Craig
College of Architecture
Georgia Institute of Technology
Atlanta, GA 30332-0155
Email: gt0183a@prism.gatech.edu

Professor Nigel Cross
Department of Design and Innovation, Faculty of Technology,
The Open University,
Milton Keynes MK7 6AA, UK
Email: n.g.cross@open.ac.uk

Charles M. Eastman
Colleges of Architecture and Computing
and Edutech Institute
Georgia Institute of Technology,
Atlanta GA. 30332-0155
Email: chuck.eastman@arch.gatech.edu

Vinod Goel
Dept. of Psychology
York University
4700 Keele St., Toronto, Ontario
Canada M3J 1P3
Email: vgoel@yorku.ca

Gabriela Goldschmidt
Faculty of Architecture and Town Planning
Technion – Israel Institute of Technology
Haifa 32000, Israel
Email: gabig@tx.technion.ac.il

W. Michael McCracken
College of Computing
and Edutech Institute
Georgia Institute of Technology,
Atlanta GA. 30332--0280
Email: mike@cc.gatech.edu

Wendy C. Newstetter Ph.D
College of Computing
and Edutech Institute
Georgia Institute of Technology,
Atlanta GA. 30332--0280
Email: wendy@cc.gatech.edu

Rivka Oxman
Faculty of Architecture and Town Planning
Technion – Israel Institute of Technology
Haifa 32000, Israel
Email: arro01@techunix.technion.ac.il

Jennifer Turns
Center for Engineering Learning and Teaching
Industrial Engineering
University of Washington
Seattle, WA 98195
Email: jturns@engr.washington.edu

Craig M. Zimring
College of Architecture
Georgia Institute of Technology
Atlanta, GA 30332
Email: craig.zimring@arch.gatech.edu

Table of Contents

Chapter 1

Introduction: Bringing Design Knowing and Learning Together

Wendy C. Newstetter
Charles M. Eastman
W. Michael McCracken
College of Computing
Georgia Institute of Technology
Atlanta, GA

1 Introduction

The need for innovative designers has never been stronger. Industrial organizations and institutions of higher education alike recognize that as society advances, design problems increase in complexity and the kinds of expertise and experience required to create effective solutions increase as well. We are continuously learning new effects of design on our health and well-being, on the technologies supporting our future, and on our culture. Over the last decade efforts to support and enhance the practices of design professionals have evolved into what some have called a *science of design*. Finger and Dixon have grouped the resulting research into six related research areas (Finger and Dixon 83): 1) discovering and describing design processes; 2) developing prescriptive models of design activity based on best practices found in industry; 3) creating computer models of design processes; 4) developing languages and representations to support design; 5) creating analytic tools to support design; 6) developing design practices for manufacturing and life cycles. These endeavors have introduced new tools supporting design and have helped practicing designers improve their processes. It is unclear, however, what contributions these six areas have made towards advancing our understanding of design learning—how engineers, architects and software and product designers—learn the skills and knowledge to be productive and innovative.

Some design educators have tried to build on this work by advocating "guided design" as a pedagogical strategy (Wales and Stager 77). Guided design prescribes the processes that good design is to follow, providing a procedural roadmap for aspiring designers. This intuitively seems a useful method for teaching novice designers, providing a scaffold for their experience and efforts. However, preliminary empirical research does not necessarily support its effectiveness. In a small controlled study with freshman designers, Atman and Bursic (96) failed to show a statistically significant difference between a group of five who had read a chapter from a design textbook and those who had not. In an ethnographic study of student mechanical engineering design teams, Newstetter (98) discovered that while students will overtly appear to follow prescriptive methods for a grade, covertly they discard the methods and undertake design as they see fit. This seems to be because they failed to see the relevance of the tools and methods provided by the instructor. These preliminary studies suggest that access to prescriptive design methods may not be as effective as a scaffolding device to learners of design as might be hoped. What these studies do not adequately explain, however, is why these potential aids to design learning failed. This is an example of the gap between pedagogical theory regarding design learning and strong links into the science of design; such links have yet to be forged.

In an effort to assess the status of current thinking with regard to design learning and its relation to other work in a science of design, the editors of this book sponsored a small conference at the Georgia Institute of Technology in the spring of 1999. A group of design researchers and educators were invited to campus to discuss an agenda for the development of a concerted research effort in design learning. The organizers (and editors of this book) sought to replicate the intellectual fervor and intensity of ideas that characterized the early cognitive science seminars, orchestrated by Herbert Simon at Carnegie Mellon University, in the early 1970's. To that end, we brought together people from different design fields that ordinarily would have little or no occasion to meet and exchange ideas. Architects, material science engineers, industrial designers, chemical engineers, software engineers and cognitive scientists, all deeply involved in design education pedagogy and curriculum issues, spent two days presenting position papers and exploring each others ideas. We envisioned this being the first of a series of bi-annual meetings that would initially forge and continually update a research agenda aimed at understanding how to better educate future designers. This collection of papers was commissioned as a result of that inaugural meeting.

Prior to the conference, the editors of this book had been hopeful that researchers from varied disciplines had already undertaken the needed foundational work. In order to identify and bring these efforts together in a journal format, we co-edited a special 1999 issue of *Design Studies*, which focused on design learning. Although we received more than twenty-five potential papers, we were disappointed with the submissions. Generally we found two problems. While many reported on innovative classroom techniques or interventions, few had conducted rigorous evaluations of learning, which would have indicated whether the interventions had been successful. It was unclear what had been learned and how novice understanding of design processes had changed. In addition, the suggested pedagogic strategies did not demonstrate grounding in the research conducted over the last two decades in cognitive science and educational psychology. As a result, we found that the design rationale for the classroom interventions was either poorly documented or missing altogether. We failed to see connections between the proposed pedagogic strategies, the learning sciences, and studies of the skills and capabilities of expert designers. It was clear from this experience that a concerted effort was needed to bring the foundational activities together in a systematic way. More specifically, bringing together means to us to identify core work, identify relevant research paradigms, and identify potential research programs. This book represents an initial step towards forging those links so as develop a *science of design learning*.

2 Organization of the book

Organizing and ordering the papers collected here have been a challenging design problem in its own right. Coming from different disciplines and communities of practice, the authors and their work represent a broad spectrum of cognitive science-based approaches to understanding design knowing and learning. The intention when we invited the conference participants was to achieve a wide diversity of opinion and perspective, a laudable goal for spurring discussion but problematic when compiling an edited book. In order to achieve a sense of coherence, we asked the authors to make changes from the original papers and solicited new papers to fill perceived gaps. Of the twelve papers, ten originally served as the basis for conference talks and appeared as drafts in the conference proceedings. The Craig and Zimring papers were solicited later from two conference participants who were not presenters.

Gaining access to the cognitive processes, situational constructs and to the knowledge that comprises expert design activity poses a considerable methodological challenge. How can we develop experimental or experiential constructs and organize data collection methods that will reveal how designers solve design problems? The designers may be at various stages of development, beginners, intermediates or experts. Where should we focus our empirical lens—on the internal workings of the mind or the external artifacts that anchor and filter cognition? How can we make the internal external and how can we see what happens to the external when it moves to the internal processing of a designer's cognition? How should we treat the intermediate products of design, the jottings, the sketches, the seemingly nonsensical trappings of the mind at work? These are questions that have dogged design researchers over the last thirty years. The chapter by David Craig provides a valuable review. Craig's paper lays out four predominant research methods that have been employed by researchers to understand how designers solve design problems. These include protocol studies, content analysis studies, process isolation studies and situated studies. Craig describes each research strategy, the kinds of data resulting from the procedure and the inherent weaknesses or questions left unanswered when utilizing such an approach. Taken as a whole, the paper acts as a primer on methodologies for studying design activity at any level of expertise: beginner, newly trained or expert. Later chapters review the various work in design learning resulting from application of one of these research methods.

The next two chapters, by Atman/Turns and McCracken/Newstetter review studies that assess student progress with design, beyond examining the quality of the product. Atman/Turns used protocol studies to assess the development of student processes during small design tasks. Their chapter reviews four such studies assessing the practices of engineering design students as they progressed through their university education. The studies relied on process measures for assessing engineering design capabilities that included alternatives generated, number of criteria considered, information categories covered and transitions. They note the general improvement of students in employing these processes during their college education. They also note significant variation among beginning students when assessed on the above process measures. Their chapter ends with pointed questions regarding the educational interventions needed to enhance student capabilities.

Most studies of design, like those reviewed in the Atman/Turns chapter, assume a strong correlation between process and product. The desired process has been identified partly through protocol and case studies and also defined more prescriptively, based on the evolving design methods and design science literature (Pahl and Bietz 96). However, the strong relation between process and product is open to serious questioning. Recent studies by (Günther and Ehrlenspiel 99; von der Weth 99) point out that experienced designers often shortcut good design practices. Such results are generally disconcerting. However, experienced designers seem to effectively replace procedural knowledge with declarative knowledge that can collapse a rigorous examination into a simple "jump" directly to a solution. The assumption about the process-product correspondence is much more complex than is usually assumed. Current assumptions are open to challenge and must be studied further.

Studying novice designers is the subject of the McCracken/Newstetter paper. They show that before-after longitudinal studies do not sufficiently address important contextual aspects of a student's beginning encounters with design. They report on two years of empirical observation in the Design Learning Lab at Georgia Tech and a large-scale survey conducted in an introductory computer science course. They discuss the challenges of collecting data on novices' conceptions of design, on student's perception of design processes and on their learning outcomes. They conclude by proposing future studies needed to better understand the misconceptions that novice designers bring to design classrooms.

A background issue in design learning is what is common across all design fields. What might be taught in a generic course of design? What aspects might be transferred from one design domain to another? Three chapters address this question. First, Nigel Cross reviews the past thirty years of work using protocol analysis to assess our current understanding of design. Based on these studies, he proposes some generic strategies used in the processes of a wide variety of design domains. He organizes these distinctive processes in terms of problem formulation, solution generation and process strategy.

Akin's chapter discusses the invariant aspects of design in relation to those aspects of architectural design that he considers variant, or unique to architecture. He identifies an architect's use of a wide variety of *analogue* graphical representations as a distinction, suggesting that these representations abstract less from the design, whereas *symbolic* representations emphasized in engineering abstract a fixed set of variables to consider. Architects also rely on *naïve* representations based on sketches and diagrams for much of their work, partly because architecture is socially situated and must respond to a variety of non-specialist stakeholders. Architects also emphasize breadth first searches of design spaces that emphasize innovation over optimized performance. Akin ends by questioning if there is a common framework of design knowledge, given the highly varied situated contexts in which different design domains operate. He suggests that this question applies within architecture, because of its constant search for stylistic innovation, as well as between it and other domains.

Zimring and Craig's chapter asks the question "what aspects of design knowledge is generic and independent of domain knowledge"? They consider various procedural aspects of designing, including ill-definedness, abduction, wicked problem solving, and design as construction. They identify counter-examples of activities that incorporate each of these activities that we would not consider design, thus raising the question whether there are any high-level processes that are truly unique to design. They end by proposing that middle-level processes, such as analogy, coherence seeking, mental simulation, dynamic modeling, argumentation, and decision making may more uniquely define design.

Identifying design's unique processes is attractive pedagogically because it suggests that there are some processes that if taught well would address the core goals of design education. However, the results of these surveys strongly suggest that there are no unique processes. Rather "design" is a socially defined activity that involves a mixture of more broadly used mental processes. It is the combining of those processes and the strategic skill in coordinating their application that seems to result in design expertise. Of course, these chapters can be interpreted in multiple ways.

If we are to understand design deeply, studies must not only map out the overall processes, but also study those processes deeply. The next four chapters attempt to examine in some depth specific processes thought to be important in design. These efforts take various positions with regard to whether the process being studied is unique to design.

The chapter by Eastman poses new questions for the study of design cognition. He proposes that the field needs to move beyond *what* designers do to *how* they do it. He poses two "how" questions. First, how are concepts that make up a design response formed? Eastman proposes that this issue may be understood by study of the relation between external representations and mental representations, how external representations are internalized and automatized. He reviews recent work in this area and outlines possible research studies. The second "how" question is, if ill-structuredness is a central aspect of design activity, how do expert designers conceptualize the design context? Here, he reviews related work in psychology dealing with memory structures and recall processes, especially those involving analogy, that may provide a general background for addressing these questions. He then surveys three research studies in design cognition that provide new understanding of how designers gain and use their knowledge in structuring the design context.

Goldschmidt's chapter follows in a manner that takes up some of the issues raised by Eastman. She reports on previous work in mental imagery and analogical mapping, both in psychology and in the design cognition literature. She proposes that visual analogy, is a basic activity used to structure and re-structure design concepts. She reviews a recent Ph.D. thesis that studied the use of visual analogy among three different levels of architectural designers. The study suggests that two kinds of skills distinguish beginners from professional designers: domain-specific declarative knowledge that allows easy solution to the many well-defined parts of a major design, and also the ability to recognize and adapt ideas to new contexts to solve ill-defined design problems. Visual analogy seems to be a central process for adapting form concepts in innovative ways.

Vinod Goel has uniquely, to our knowledge, begun to relate the processes of design cognition to its neuroscience basis. In this chapter he reports on a patient PF, an experienced architect, who experienced a right hemisphere lesion of the prefrontal cortex. This impairment provided a unique opportunity to study the contribution of the prefrontal cortex on design behavior. He reports on PF's ability (or inability) to deal with two different kinds of design information: (1) the domain specific issues, forms and materials used in defining and solving design problems, and (2) the tacit planning and structuring processes that are needed to resolve large ill-structured design tasks. While Goel could not undertake a before-and-after study of PF's disability, he instead matched him with a control subject of similar background and skill. By tracking both PF and the control undertaking the same design task, Goel shows that the physical assault on PF's cortex led to marked inabilities to deal with the ill-structured aspects of design. While PF was well able to carry out the needed recall of criteria and issues, identifying forms and materials that make up a solution and other domain-specific declarative types of knowledge, he was not able to adequately plan or execute the detailed activities of matching criteria to actions, hierarchically organizing actions or other planning activities needed to generate a working solution. The study points out critical skills that, if missing, prohibit effective design behavior.

Carroll's orientation is software design. His chapter reviews the work on and rationale for scenario-based design. Here, scenario-based design is the reliance on embedding design specifications and other planning materials in a task-related context. He reviews why scenario-based design specifications provide an important contextual perspective necessary for good design results. He also reviews their use in documentation and testing. The benefits of scenario-based design that Carroll reviews include the support for reflection, relaxed commitment, multiplicity of uses and their ability to be re-used. Scenario-based design responds to the issue of how to capture and communicate the context and situated aspects of a design task to designers that have no involvement in the context where a design is to be applied.

These four studies begin to address the middle level activities identified by Zimring and Craig as possibly providing a core set of generic activities comprising design. Yet a careful review suggests that some of the activities studied are rather domain-specific. Visual analogy, as presented by Goldschmidt,, addresses spatial composition issues encountered in architecture; scenario-based design is important in software engineering because of the social structure of software engineering, where implementers are distinct from requirements generators and other actors in a highly vertical process. On the other hand, Goel's study of planning and mental integration incapacities associated with the right prefrontal cortex is probably common to a wide variety of design and non-design tasks. It would be interesting to learn if PF's impairment also applied to such tasks as planning a vacation or filling out his income tax.

The ultimate test of studies of design learning is the translation of concepts and empirical results into classroom exercises. The last collection of papers report on efforts to transform the pedagogy of design learning in classrooms. The two efforts reported are based on quite different conceptualizations of the central issues in effective design. Each author draws upon a different research base for his or her educational proposals.

Oxman opens with a review and critique of traditional design education frameworks pointing out that the measure of student learning is generally equated with the quality of the designed product rather than on the learning that has occurred as a result. She argues that a central subject of design education is the development of knowledge structures about the designed artifact. She proposes that the construction of these knowledge structures is enhanced if they are defined and modeled explicitly, while designing. She argues that they facilitate explication of various reasoning processes that are difficult to address if not made explicit. The processes include concept development, abstraction, refinement and mapping between representations. She recounts experiments in which students were asked to construct knowledge networks of their designs using computer-based diagrammatic models, as a pedagogic technique for fostering design understanding.

The ineffectiveness of much of current university-based design education is decried by Bucciarelli. He outlines the weaknesses resulting from the lack of context and situatedness that arises from the specialization and limited focus that is embedded in most of engineering. He proposes a different framework for defining analysis as well as design tasks that embeds them in a richer context that allows students to deal with the ill-definedness of detail objectives, criteria and solutions that are part of real-world design.

Both of these examples of curricular development draw directly from the research literature for their rationale for improving design education. Oxman draws from the cognitive psychology and AI literature regarding knowledge structures, while Bucciarelli draws from the situated learning literature. Neither provide an assessment of the costs or benefits of their pedagogical approach.

3 Summary

The groupings of chapters in this volume tacitly suggest a structure for considering future work in a science of design learning.

Methodology: All work must be founded on clear methodological and paradigmatic assumptions. That is, all work that studies design assumes an underlying model of the process; the model identifies the activities or variables of importance. All studies also must deal with issues of data collection, its reliability and generality. The ill-defined problem solving paradigm is still the most widely used and can probably be used to learn much more about design. It competes with anthropologically-based contextual studies that grounds activities not on individual cognition, but rather on the framework of activities defined and embedded in a group or culture. Both types of study rely on a small toolkit of behavioral science data collection methods, surveyed by Craig. Methodological issues form one cornerstone of any science.

Longitudinal Assessment of Design Learning: The chapter by Atman and Turns and also the chapter by Goldschmidt carry out studies that assess beginning and advance designers, identifying changes in the behaviors of beginners in comparison to seasoned experts. The Atman/Turns chapter is explicitly focused on assessing overall design skill, where the assessment is made according to the activities embedded in the tested student's design process. Such studies are the ultimate assessment of the impact of a design education. However, we have only begun to seriously address the issues implied by such studies. All assessments of design capability are based on certain assumptions—about the relation of process to product, about the goal of design education and about the character of good design. These assumptions are open to question, as we have noted earlier.

Assessment of Beginners, Experts and Special Populations: Most studies have focused on the processes and knowledge structures of knowledgeable designers. The chapter by

Cross and the introduction in the Eastman chapter are example summaries of what we have learned about experts. The chapter by McCracken and Newstetter challenges us to recognize that the starting point of beginning designers is often filled with misconceptions. Unlearning and destruction of these misconceptions may be needed. Goel's chapter shows the value of recognizing the potential insights that can be gained by carrying out design cognition studies on special populations. By taking advantage of situations where the state of normal expert designers change, or where unusual specialized training may have occurred, or by observing how children develop both good and bad conceptions about design processes, we may better understand the context of design learning and the mental context that supports it.

Important Component Processes: The middle collection of papers present examples of work focusing on important component process of design. Four types of component are considered in some detail. Goldschmidt's chapter is a clear example, focusing on visual imagery. Eastman's chapter reviews work in two related areas, those dealing with design representations—both externally taught representations and internal mental representations— and with design knowledge structuring and recall. Here he also focuses on analogy, but with a wider consideration of mental adaptation and transfer. Carroll's chapter argues the importance of an external representation for better capturing the contextualism and situatedness of software design problems, though he does not use this terminology.

Structure of Design Knowledge: The general assumption by most researchers in design cognition is that high-level design processes—especially those used at the initial stages to conceptualize and structure a design task and the high level communication and coordination activities —are general and shared across different domains of design. It is the detailed domain-specific content in chemical engineering, material science, architecture, structural and geophysical engineering that are unique to a design domain. Also, the processes developed for evaluating and possibly integrating and optimizing these domain-specific issues are often specialized to specific design fields. Akin's chapter reviews what he considers domain-specific in architecture, while Cross's chapter surveys general design procedures that he has culled from design cognition studies using protocol analysis. The Zimring/Craig chapter challenges all design learning researchers to consider if there are any processes that are unique to design. They review four different proposals and reject them all, suggesting that design may better be considered as a broadly defined cultural domain that relies on general but not unique problem solving skills.

Design Cognition in the Classroom: The outcome of a science of design learning must be curriculum designs that are based on the knowledge gained about the state of students learning design and effective interventions that add to their knowledge and processes. None of the authors here make strong proposals, with a strong theoretical foundation. However, the Chapter by Bucciarelli makes a plea for a more situated approach for education in engineering design, as one facet of a design curriculum. Oxman comes closer to developing a curriculum based on design cognition theory and describes its implementation in a course she has developed. It attempts to make explicit for design students the structure of knowledge they are developing. It provides them with a conceptual framework for discussing knowledge processing issues. Some may question its seeming emphasis on further abstracting student's experience. We look forward to objective assessments of this pioneering effort.

4 Acknowledgements

Many people and organizations contributed to this project. We are grateful for the financial support of the EduTech Institute, the College of Architecture and the College of Computing at Georgia Tech. The Proctor and Gamble Foundation provided financial support to EduTech for some of our early work in design learning. We thank the Dean of the College of Architecture, Tom Galloway and the Dean of the College of Computing, Peter Freeman, who have created an intellectual environment that recognizes the importance of design and design learning.

We are also grateful to all of the conference participants, including those who did not present papers at the conference. The active dialogue of all of the participants improved the contents of this book. In particular, we would like to thank the moderators of the sections of the conference: Ashok Goel, Mark Guzdial, Janet Kolodner, Colin Potts, and Craig Zimring, who all led interesting discussions about the papers at the end their session. One person we would like to thank personally is Paul Feltovich for his insightful comments at the conference.

Any book such as this has an editor that sweats the details. Our editor, Jennifer G. Sheridan, did much more than sweat the details. We feel her work has insured the quality of the product. As is always stated, if it's right tell her, if it's wrong tell us.

Thanks as well to Saif-ul Haq who designed the cover for the book.

Finally, the people who wrote the papers in this book deserve the most credit. Their contributions are what make this a book. We want to thank each of them for their efforts, and most importantly their intellectual contributions to the field of design learning.

References

Atman C J, and Bursic K M, (1996), Teaching engineering design: Can reading a textbook make a difference? *Research in Engineering Design,* 7(7), 240-250.

Bucciarelli L, (1994), *Designing engineers*, MIT Press, Cambridge, MA.

Chi M T, and Ceci S J, (1987), Content knowledge: Its role, representation, and restructuring in memory development, H. W. Reese Ed. *Advances in child development and behavior*, Orlando, FL, Academic Press Vol. 20, 91-142.

Chi M T, Glaser R, and Farr M, (1988), *The nature of expertise*, Lawrence, Erlbaum and Associates, Hillsdale, NJ.

Clement J, (1982), Students' preconceptions in introductory mechanics, *American Journal of Physic* (50), 66-71.

Driver R, (1983). *The pupil as scientist?* Milton Keynes, Open University Press.

Finger S, and Dixon J, (1983), A review of research in mechanical engineering design. *Research in Engineering Design,* 1(1).

Günther J, and Ehrlenspiel K, (1999), Comparing designers from practice and designers with systematic design education, *Design Studies*, 20(5), 439-451.

Glaser R, (1984), Education and thinking: The role of knowledge, *American Psychologist, 46*, 1-46.

Glaser R, and De Corte E, (1992), Preface, F. J. R. C. Dochy Ed. *Assessment of prior knowledge as a determinant for future learning*, B.V./Jessica Kingsley Publishers, Utrecht, London, Lemma, 1-2.

Johnson-Laird P N, (1983), *Mental Models*, Harvard University Press, Cambridge, MA.

McClosky M, (1983), Naive theories of motion, D. Gentner and A. L. Stevens Eds. *Mental models*, Lawrence Erlbaum and Associates, Hillsdale, NJ.

Newstetter W, (1998), Of green monkeys and failed affordances: A case study of a mechanical engineering design course, *Research in Engineering Design, 10*, 118-128.

Pahl G, and Beitz W, (1996), *Engineering design*, Springer-Verlag, London.

Resnick L B, (1983), Mathematics and science learning: A new conception, *Science, 220*, 477-478.

von der Weth R, (1999), Design instinct? — the development of individual strategies, *Design Studies* 20(3), 453-463.

Wales C F, and Stager R A, (1977), *Guided design*, Morgantown, WV: Center for Guided Design, University of West Virginia.

Chapter 2

Stalking Homo Faber: A Comparison of Research Strategies for Studying Design Behavior

David Latch Craig
College of Architecture
Georgia Institute of Technology
Atlanta, GA

Abstract

Four different strategies for studying design behavior are examined and compared: think-aloud protocols, content analysis, process isolation and situated studies. The four strategies span a range of data collection and analysis methods. More substantial differences, however, are found to exist in the theoretical biases each entails. Think-aloud protocols assume that design can be described as search; content analysis assumes that the structure and content of internal representations is critical to understanding problem solving behavior; process isolation assumes that design can be cognitively decomposed; and, finally, situated studies assume that a better understanding of design involves the study of designers and their environments as integrated systems. Despite their sometimes competing claims, these four strategies can potentially work together in complementary ways to provide insights into design behavior.

1 Introduction

How do designers solve design problems? What exactly is it that designers do? Researchers in a variety of fields have sought answers to these questions for a variety of different reasons. Researchers in cognitive psychology, for example, have looked at design both because it represents a core aspect of human creativity and because of the challenge it poses in applying what is already known about cognition to more complex domains. Researchers in fields like engineering, computer science and architecture, on the other hand, have looked at design because of a larger interest in improving what designers do through the use of design aids and better teaching methods. At the same time, design has been the subject of study in a host of other fields, by researchers pursuing their own discipline-specific goals, ranging from the sociological to the anthropological to the historical.

The discussion of research strategies in this chapter is not limited by disciplinary boundaries. It is, however, limited by the general goals of research. In particular, the chapter addresses empirical research strategies aimed at understanding how designers solve design problems. Despite this specific focus, a lot of room is still left for diverse strategies to be played out, particularly given the scope and complexity of the subject being studied. Design problems are not solved in any simple or unitary way. They are solved in a range of social, cultural and physical settings, through the efforts of individuals as well as groups, using a variety of different types of knowledge and skills.

In all, four research strategies are considered: think-aloud protocols, content analysis, process isolation and situated studies. Although a variety of published studies are discussed using this classification scheme, not all fit neatly within it. Most studies blend together multiple strategies at once, resulting in almost as many distinct approaches as there are studies themselves. Moreover, this classification scheme does not necessarily represent different data collection techniques or methods of analysis. The strategies differ, rather, primarily in terms of their theoretical orientations. Although some studies appear schizophrenic in their use of theory and method, theoretical orientation is noted wherever it is made explicit in published reports. In some cases, liberty has been taken in reclassifying studies to help illustrate less conspicuous research alternatives. Along these lines, one of the goals of the chapter is to help strengthen strategies that have been used by researchers but, for lack of theoretical and methodological clarity, may not have been used to their full potential.

Before getting into the research strategies themselves, two assumptions will be made, first that design is a distinct type of behavior and second, that design behavior is open to description. Although the first assumption is subject to debate, no special claims will be made about what design is at this point. Design will simply be considered any problem solving activity that results in the creation of an artifact or a plan for generating an artifact. Different research strategies, of course, have different things to say about what design is, as will become evident later in the chapter. The second assumption is somewhat less controversial. Although at times it may seem that design follows divine inspiration, that it is ineffable or that it withers under description, there are too many levels at which one can explain human behavior and too many techniques for elucidating reasoning to make any claims against empirical research outright. Not all research strategies, of course, will be effective, nor should we assume that a single all-encompassing description of design is realizable. Some strategies, however, may be more appropriate for studying design than others, and the purpose of the chapter is precisely to see how different strategies stack up.

2 Think-aloud protocols

In what is perhaps the most thorough account of how one might study problem solving behavior, Ericsson and Simon (84) present a strategy that hinges on having subjects talk aloud or think aloud while executing a task. The goal of the strategy is to track and better understand the sorts of reasoning processes and strategies employed by problem solvers. The "think-aloud" version assumes that some mapping may be required between verbalization and internal representations, whereas the "talk-aloud" version assumes that internal representations are already in verbal form. "Thinking aloud" is thus more general and will be used here to refer to both methods. Although other strategies for studying cognition and behavior require subjects to talk while performing a task, think-aloud protocols will be used to refer specifically to methods that use concurrent verbalization to track thought *processes*.

Ericsson and Simon's defense of think-aloud protocols begins with the observation that all data collected in behavioral experiments, whether verbal or not, is subject to interpretation. Hence, there is no *a priori* reason to assume that data collected in think-aloud protocols will be any less objective than data collected in other experiments. Ericsson and Simon do not assume that subjects have direct access to their mental processes, nor do they assume that thought is "inner speech." They only assume, consistent with their view of verbal reports as behavior, that one can infer upon hearing a subject say "X" that the subject was *able* to say "X." Following accepted cognitive theory, one can then infer that if the subject was able to say "X," the subject had an informationally equivalent representation of "X" active in short-term memory (STM). Finally, if one also accepts a serial information-processing view of cognition (i.e., that thought consists of "a sequence of internal states successively transformed by a series of information processes"), one can infer that "X" is somehow related to what was stated earlier and what will come later.

Ultimately, think-aloud protocols require the interpretation of data at two somewhat distinct levels of theory in order to infer underlying thought processes. The first level is low-level cognitive theory of the sort mentioned above, while the second level is the level at which hypotheses are made. The two may interact but potentially in a positive way. Ericsson and Simon refer to this as "bootstrapping": "We need a model in order to interpret data that are to be used, in turn, to test the model. Under these circumstances, our data-interpretation model should be as simple as possible, and it must not incorporate components that are themselves bones of theoretical contention" (Ericsson and Simon 84, p. 10).

The main difficulty in making use of think-aloud protocols lies with the basic task of interpreting utterances and relating them to prior utterances. This is where reliance on higher-level theories comes into play. According to Ericsson and Simon, protocol data should be coded based on a theory of what sorts of things a subject might reasonably be thinking about. In the wealth of protocol studies conducted by Newell and Simon (72), for example, this entails mapping out plausible problem spaces that subjects might traverse in the course of completing a task. Plausible problem spaces, in turn, yield a list of concepts that will be relevant, which then serves as a "dictionary" for interpreting utterances. The *a priori* formalization of an encoding will, of course, help guide an interpretation even if the definitions are open-ended. For example, Ericsson and Simon point out that a particular mapping from linguistic utterances to concepts might result in identical translations for "red circles are zog" and "blood-colored circles are zog." "Blood-colored" can be assumed to contain the same information as "red" as long as one's high-level theory of what the subject is thinking does not count the two as being different.

Think-aloud protocols appear to work reasonably well when studying how subjects solve well-defined problems like the Tower of Hanoi and cryptarithmetic puzzles. In such cases, the way the problem is presented to subjects leaves little freedom in representing the space of possible moves and solutions. Thus, "whenever verbalizations correspond to plausible intermediate states in a processing model for the problem solving activity, we can plausibly infer that this information is actually used in generating the problem solution" (Ericsson and Simon 84, p. 171). From there, one can generate hypotheses about how a subject might navigate or transform a problem space (e.g. what sort of heuristics will be used) and compare these with how a subject actually traverses or transforms such a space.

One criticism of think-aloud protocols is that although verbal data can be used for a variety of purposes, it may be inappropriate for tracking sequential thought processes. In particular, think-aloud protocols may misrepresent underlying processes to the extent that subjects "sometimes cannot report on the existence of critical stimuli, sometimes cannot report on the existence of their responses, and sometimes cannot even report that an inferential process of any kind has occurred" (Nisbett and Wilson 77, p. 233). Ericsson and Simon respond to such attacks by arguing that it is only in experiments where subjects are explicitly asked to explain *why* they formed certain judgments or behaved in a certain way that fallacious reports are generated. This, they point out, is not consistent with their proposed use of verbal data. Ericsson and Simon never claim that subjects have access to cognitive processes; at best, they claim subject can report the contents of STM. In Nisbett and Wilson's study, all accounts are retrospective, in which case it is possible that traces of earlier processing have disappeared and must therefore be inferred. Ericsson and Simon specifically note the potential dangers of retrospection: subjects may retrieve the wrong episode, they may retrieve information that is related to but not part of a prior episode, they may simply infer a plausible course of reasoning, or, finally, they may unconsciously neaten up an episode, leaving out missteps and reasoning errors that were actually part of their reasoning process.

Other criticisms of think-aloud protocols are that verbalization provides an incomplete account of the reasoning undertaken by subjects even if verbal reports are accurate, and that verbalization may potentially interfere with thinking processes in other modalities. Both of these criticisms are limited if one is strictly dealing with well-defined tasks. If, in other words, a subject is given enough information to clearly specify a problem space, gaps in the protocol may be interpolated since, as evidenced by a task analysis, only a limited set of relevant information states can be instantiated. If a subject's thoughts wander to last night's baseball scores while solving a well-defined puzzle, we can rest assured that those thoughts will not affect the puzzle solving process in any qualitative way. In the same sense, verbalization is less likely to interact with reasoning in other modalities if one is applying stepwise procedures towards achieving a goal (Schooler and Melcher 95). Verbalization, if anything, may help subjects reflect on the state they are in and develop a more explicit map of a well-defined space.

The incompleteness of think-aloud protocols may be more problematic, however, if the problem space is not well defined. In such cases, the experimenter must simultaneously infer both the problem space and the state the subject is in from the verbal data. Ericsson and Simon recognize this problem, noting that in actuality problem spaces associated with even the most well-defined tasks can change unpredictably over time. When faced with an unclear problem space, Ericsson and Simon suggest using the weaker criteria of relevance and sequential consistency to establish the contents of and links between information states. They add, however, that these criteria preclude the study of "relatively unconstrained cognitive activities like free association or daydreaming" (Ericsson and Simon 84, p. 171), as well as processes that are not "cumulative." In this case, despite Ericsson and Simon's general attempt to rescue think-aloud protocols from problem space uncertainty, problems may still persist in coding design protocols. For one thing, design may not involve strictly cumulative cognitive processes. A designer's thoughts may wander laterally, and free-association and daydreaming may, in fact, be useful. This, of course, amplifies the problem of inferring what a designer is thinking and how it fits in with a longer train of thought. One might rely strictly on the local context of utterances to infer the knowledge state of a designer, as suggested by Dorst and Dijkhuis (96), but such an approach seems to trivialize the nature of internal representations. Internal representations are potentially complex, so much so that verbalization may only reveal the tip of the representational iceberg (see e.g. Barsalou 93).

In addition to incompleteness, interference with other modalities may also create problems for think-aloud protocols in ill-defined domains or, for that matter, any domain where subjects are not able to make use of strategies that correlate well with speech. Think-aloud instructions in insight problems have, in fact, been shown to hamper problem solving ability, beyond simply slowing down the completion of a task (Schooler and Melcher 95). Although it is not clear what this says about the underlying reasoning processes, verbalization may tie up resources that could be used in other ways, or it may bias particular (e.g. logical) reasoning strategies over others. Among other things, experiments have shown that verbal descriptions of visual imagery impair memory for such scenes (Schooler and Engstler-Schooler 90; but see Freund, cited in Loftus and Bell 75, for different results). Hence one might assume that, at a minimum, verbalization will limit a problem solver's ability to take full advantage of visual information.

The next section reviews the use of think-aloud protocols in design and addresses their potential limitations in more detail. Although, as mentioned earlier, there are a variety of research strategies that require subjects to talk while performing a task, the studies referenced here all make use of Ericsson and Simon's proposed strategy for using think-aloud protocols.

2.1 Think-aloud protocols in design

Think-aloud protocols have been employed in design studies for more than thirty years. In that time, a variety of distinct approaches have been taken. All of the studies referred to in this section describe design behavior as movement within a problem space and use protocol data either to illustrate, support or expand cognitive models of design. To help organize the discussion, the studies that are mentioned are grouped into two categories: those that describe search in terms of a set of operators or production rules for generating moves, and those that treat search more generally, using quantitative analysis rather than model instantiation to identify patterns of behavior.

2.1.1 Production-rule approaches

The earliest application of think-aloud protocols in design (Eastman 69) and some later applications (e.g. Akin 78, Ullman et al. 88) model the design process with a set of cognitive operators that generate and transform information about the problem or about a developing solution. The problem space, in this case, is assumed to be locally rather than globally defined, consistent with Simon's (73) description of ill-structured problem solving processes. In other words, information is assumed to be attended to and manipulated in a limited fashion, such that the problem space changes continuously throughout a problem solving episode as new information comes into view. Operators that are postulated typically deal with things like the retrieval of requirements, constraints and general knowledge from long-term memory (LTM); the assimilation of information from external sources (e.g., books); the transformation of such information through deductive inference or through the recognition of patterns; the transformation of requirements into form; and finally, the fine tuning of form variables to match constraints. In simpler terms, production-rule models argue that constraints and requirements are transformed into information that can then be used to specify forms. The latter usually lead to multiple alternatives, since form variables and subtype specifications are often left open. Using an example from Akin (78), the requirement that sleep be supported in the design of a house can be (partially) transformed into the specification of a bed, which can then be varied in terms of size, style and placement.

Design processes modeled on production rules are not necessarily deterministic. Because the space of potential design decisions can be large and because designers may not have enough expertise in a domain to follow a set decomposition, the result can be idiosyncratic. The kinds of information initially brought into the problem space and the sequence in which requirements and constraints are addressed may vary, depending on personal experiences, on the structure of information in LTM and on the retrieval cues present in the environment. As Ullman et al. (88) points out, a design process that results in a surprising or innovative design can be modeled with production rules to the extent that the decomposition need not be known in advance. Aside from this variability, however, design processes modeled using production rules are analogous to those associated with solving puzzles: some complex combination of component types and variables must ultimately be specified, usually incrementally, to satisfy a set of requirements and constraints. Although rules can also be used in design to transform requirements and constraints themselves (which is not usually the case with simple puzzles), this can be seen as simply adding an additional layer of complexity to the problem solving process.

Although think-aloud protocols have been used to illustrate production-rule models and in some cases to expose novel search strategies (e.g. Akin 78), there are several questions concerning this strategy. In particular, one might ask whether rule-based knowledge is sufficient for describing the sorts of reasoning that go on in design and, hence, whether think-aloud protocols coded using production-rule models accurately or fully capture design reasoning. Although rule-based reasoning is superficially evident in most protocols (in the form of rationale like, "I'll add a bed here *because*..."), this sort of reasoning may simply mask the use of more complex relational knowledge structures (Barsalou 93). Notably, it has been found that when individuals are put in controlled situations that trigger the retrieval of specific knowledge structures or schemas, without their knowing, they often manufacture rationale after the fact to explain subsequent judgments simply because they have no other way to account for their reasoning (Nisbett and Ross 80). Their rationale, moreover, may not be consistent with their actions. Although, as mentioned earlier, such research relies on error-prone retrospective accounts, the issue here is not simply whether verbal reports are correct or incorrect. The point, rather, is that reasoning may be buried in extended knowledge structures that do not surface in the form of isolable, easily verbalized rules.

The primary difference between reasoning from schemas and reasoning from rules is that schemas carry with them a halo of unconscious, but often consequential, assumptions. If one reasons about a particular person, for example, one might reason from a stereotype and actually infer features, or traits, that are not present. In design, the complexity and variety of retrieved and assimilated information suggests that a lot of unconscious assumptions are folded into the problem solving process. Some unconscious assumptions amount to little more than the instantiation of default values in knowledge structures. For example, one might add a bed to a room and assume that it is white without consciously considering color. Designers might also, however, invoke knowledge structures from analogous domains, in which case conceptual constraints and structural relations may direct the development of a solution unconsciously. In Ullman et al. (88), for example, a designer working on the design of a mechanical device assumes early on that one component will be used to satisfy two crucial functions, only to overcome this self-imposed restriction late in the design process. The subject, in this case, appears to have retrieved and utilized a schema for a particular kind of mechanism without actually considering the structure of the retrieved representation or the fact that a specific schema was retrieved at all. The representation, in other words, guided the subject's behavior without the subject consciously making any decisions.

The use of think-aloud protocols for illustrating, testing and exploring production-rule models of design comes down to the ability of the experimenter to infer a potentially large and complex set of connections between utterances made by subjects. This requires empathy, since the problem space is not made explicit in the problem statement but rather develops out of the designer's actions. As Eastman points out, "the analyst can understand the problem solver's processes to the degree that he can find correspondences between the processes he has experienced and thus understands, and those of the subject" (Eastman 69, p. 670). The problem, in this case, is that even if the experimenter is empathetic, he or she may simply be making the same unconscious assumptions that the subject makes and hence fail to uncover all of the underlying processes. Given this limitation, it is tempting to describe the use of think-aloud protocols to model processes in design as "concurrent introspection." There are, of course, design domains in which expert knowledge is well codified, and there are also design problems for which components are completely (or largely) given in advance and for which variables and subtype specifications are limited. Both would potentially make the task of inferring thought processes from think-aloud protocols more plausible. This, for example, may be the case in Eastman's (69) study of the design of bathrooms given a fixed budget and a fixed enclosure. However, even in the case of highly constrained bathrooms, everyday concepts about bathroom use, which may carry a host of unanalyzed assumptions, may enter into the process.

2.1.2 Quantitative approaches

Other design studies that utilize think-aloud protocols of the sort described in Ericsson and Simon (84) have taken a much looser approach to analyzing behavior. In particular, some studies have taken a quantitative approach, where utterances are segmented, categorized and analyzed for correlations, frequencies and patterns (e.g. Akin and Lin 96; Gero and McNeill 98; Goel 95; Goldschmidt 96, 97; Lloyd and Scott 94). Such studies rarely concentrate on complete cognitive models of design. Instead, they begin with intuitive ideas about things like the fomenting of critical decisions or the behavioral underpinnings of effective problem solving behavior, and then use pattern analysis to fill in the details. Although starting off without well-defined hypotheses may suggest that this strategy is vulnerable to confirmation bias; in practice it is analogous to an archeologist combing through layer upon layer of sedimentation. Hypotheses that are specific to the data must take shape as the data becomes available simply because data for answering specific questions cannot be generated willfully. In this spirit, quantitative approaches are less likely to be called experiments and more likely to be called "macroanalysis" or "measuring designing" (Gero and McNeill 98).

Although the quantitative studies discussed in this section focus on the structure of the design process, they nonetheless employ search space models of problem solving at the outset and frequently refer to Ericsson and Simon (84) for methodological support. They differ from the studies discussed in the previous section, though, in that they look for statistical correlations and patterns within verbal reports, sometimes on the assumption that the design process is too unstructured to be tracked using production-rule models. In this sense, they acknowledge that subjective and often unconscious knowledge structures wreak havoc on detailed analyses and thus drop content altogether. In the case of Goldschmidt (96), for example, every coherent statement made by the designer is assumed to constitute a move that can be linked to other moves in unpredictable (both to the designer and to the experimenter) ways. Links are not coded as a sequence of moves simply because it is considered too difficult to say where they came from.

The problem with maintaining a problem-space model of design while focusing on statistical correlations and patterns within utterances is that it can be difficult to tell exactly what a move, information state or operator is. In traditional problem space studies (e.g. Newell and Simon 72), a state in a problem space is pinpointed by actually tracing a sequence of moves. In chess, for example, a potential state is one that can exist in a legal game. Certain configurations of chess pieces on a chess board are simply impossible. By taking a quantitative approach, however, all evidence of a tractable problem space is blurred, making the assessment of a given utterance especially difficult. Encoding a designer's utterances as evidence of movement through an unknown problem space seems unfortunately like trying to follow the movement of a mouse through an invisible maze to better understand the mouse's behavior. In this case, every twitch could be read as a purposeful move and vice versa.

A second concern with the quantitative analysis of think-aloud protocols is that frequencies, correlations and patterns do always help to clarify causal links. Aside from the intuitive theories that accompany such studies, little is ventured about the cognitive processes or mental representations that might underlie reasoning in design. Patterns or correlations found in protocol data could simply reflect the conceptual bias of the coding scheme or, at best, indicate a relationship between behaviors without suggesting how they are related. Quantitative analysis is generally useful when comparing two refutable hypotheses. This, however, is not usually the case in design studies. Most hypotheses deal indirectly with the data. Goldschmidt (96), for example, postulates that "effective design processes are characterized by a high ratio of interlinkings among its moves" (p. 74). Her study, however, does not look at effectiveness but rather at how "link densities" and other patterns compare between individuals and groups. As another example, Akin and Lin (96) speculate on the significance of multi-modal behavior (e.g. drawing, examining and thinking at the same time) and find that it is more likely to coincide with "novel design decisions" than with routine decisions. They concede, though, that their analysis does not suggest which fosters the other or if, in fact, they are causally related at all.

A final concern with quantitative analyses involves the nature of the coding schemes. As mentioned above, coding schemes are created to partition utterances into categories of behavior, usually at a level of abstraction above production-rule models. Once a coding scheme is created, agreement between independent coders is often measured to see how sharp the scheme is. One problem is that lack of agreement – as low as 60% in some studies (see e.g. Lloyd and Scott 94) – does not always deter researchers from using a particular coding scheme. A more insidious problem, though, is that the definition of a coding scheme may be conceptually tenuous, confounding findings produced by it. As one example, Goel (95) uses a quantitative analysis to comparatively study movement between design "modules" (which he defines as distinct design components or design issues) in well-defined and ill-defined tasks. The question, in this case, is what the modules will be for a given task. Goel makes the assumption that in cryptarithmetic problems (which require subjects to decode letters into numbers in order to satisfy an arithmetic expression, such as SEND+MORE=MONEY), the columns of the arithmetic expression, as opposed to the individual letters or letter types, are modules. This coding assumption directly impacts Goel's findings, which deal with the frequency of moves between modules. Although the assumption is contestable, it gets little theoretical support in the study. Following Ericsson and Simon's (84) discussion of the think-aloud protocols, utilizing theoretically unclear coding schemes may lead to "bootstrapping" failures and ultimately jeopardize the results of a study.

3 Content Analysis

The notion of relational knowledge structures, or schemas, that Nisbett and Ross (80) invoke to explain implicit inferential behavior has developed over the past several decades into a fairly detailed theory of conceptual structure and conceptual reasoning. The basic idea is that concepts get their meaning from links with other concepts, often inheriting or sharing whole complexes of links with other domains, near and far. Links may be recursive (e.g. *humans* have *progenitors* which are also *humans*) and may be constrained in complex ways (e.g. a *ski vacation* has a *location* which must be *snowy*). Some knowledge researchers have worked to develop detailed computational structures that are capable of accounting for a wide variety of reasoning in a unified way (e.g. Minsky 75; Rumelhart and Ortony 77; Schank and Abelson 77; Barsalou 93), while others have used relational structures to analyze new experimental data, for example in conceptual combination tasks (Wisneiwski 97). Likewise, in design, schema retrieval, combination and modification has been used to describe how designers develop solutions to problems in ill-defined domains (e.g. Gero 90).

A research strategy designed specifically to highlight conceptual structure and conceptual change, as opposed to rule-based processes, is what Chi (97) refers to as "verbal analysis." While Chi's interest is primarily in education – that is, in how students' representations change over time, how they vary in different contexts and how they differ from those of an expert – her strategy for studying problem solving is also applicable to design. In this section, verbal analysis is used as a primary example of a more general strategy that will be referred to as content analysis. Unlike think-aloud protocols, content analysis does not aim to track cognitive processes, although it may rely on verbal reports for data. Whereas Newell and Simon (72), for example, perform a task analysis to deduce probable problem space representations prior to a problem solving episode, content analysis uses verbal data and other externalizations like sketches to infer representations after the fact. This is argued to be particularly relevant in ill-defined domains where problem spaces depend on more than that which is given in a problem statement. In the end, Chi argues that processes are secondary, since "it is the solver's representation that determines the problem solving process" (p. 278).

To the extent that content analysis aims to uncover the structure and content of representations, it relies heavily on the qualitative analysis of external, often verbal, reports. However, as Chi points out, quantitative analysis may be used to back up qualitative findings. More precisely, "one tabulates, counts and draws relations between the occurrences of different kinds of utterances to reduce the subjectiveness of qualitative coding" (p. 273). An example of this might be counting how many times a problem solver refers to prior problems or examples and comparing those results with likelihood of success on the current problem. Counting, in such cases, usually involves predicting some behavior based on what a subject is thought to know. Such knowledge can be also be tracked by cataloguing the content of utterances and the conceptual links between them. Important questions might, for example, be whether the conceptual structure boils down to a list of perceptual features, a set of fragmented beliefs, a set of central exemplars, or an analogous concept. Each of these may have different consequences for the type of reasoning that follows and ultimately for the way an individual solves a problem. A designer, for example, who works with a set of fragmented beliefs may utilize information erratically or inefficiently, whereas a designer who works from a well-understood analogous concept may be able to manage a variety of concerns and insights by integrating them into an already coherent conceptual structure.

Like think-aloud protocols, content analysis may involve asking subjects to say what they are thinking while solving a problem. However, since the aim of content analysis is to uncover internal representations, some probing on the part of the experimenter is usually admitted. Obviously, probing may affect performance and may alter the underlying processes. This, however, is seen as consistent with the overall goals of content analysis for two reasons. First, processes are not the primary focus, and thus may be sacrificed if necessary. Asking a designer specific questions about a problem or about existing concepts without actually having them carry out a design might still reveal what sort of knowledge they would use. Second, interfering with underlying processes as they occur may be beneficial provided that the interference can be tracked. One goal of verbal analysis is to understand how representations change through reflection; hence instigating change may lead to valuable insights.

Also like think-aloud protocols, content analysis relies on theory for the coding of external reports. Although Chi criticizes traditional think-aloud protocol studies for their lack of a control condition (as Chi points out, "because one can always tweak the model to 'match' the protocol, it's never clear how the model can ever be invalidated"), think-aloud protocols are not necessarily aimed at picking out hypotheses as much as they are at simply organizing behavior in a way that makes discovering higher-level patterns of behavior easier. Content analysis is similarly dependent on theory for telling us what to pay attention to and what to encode as part of a representation. Even if quantitative performance measures are incorporated into content analysis, some faith is required to accept that the variables being measured are sufficiently independent and significant enough to warrant studying. Thus, content analysis is not necessarily a more rigorous research strategy. Content analysis and think-aloud protocols are perhaps best considered complementary strategies, the former being tailored for ill-defined domains and investigating the formation and use of knowledge representations, the latter being tailored for well-defined domains and investigating step-wise processes.

3.1 Content analysis in design

To a certain extent, the quantitative analysis of think-aloud protocols discussed in Section 2.1.2 resembles content analysis, particularly verbal analysis. Both develop quantitative coding schemes based on preliminary qualitative analyses and work between the two reflexively. They differ, however, in that think-aloud protocols focus almost exclusively on the process of design at the expense of representational structure, while verbal analysis does the opposite. Content analysis, moreover, encourages the use of probes to elicit particular kinds of externalizations.

In practice, few design researchers have undertaken studies that delve into the structure and use of knowledge representations through verbal reports or other externalizations. In part, this may be because designers are assumed to be generalists, lacking any easily pinpointed expertise. Chi, in her own studies, has used verbal analysis in relatively well-structured domains like physics (e.g. Chi et al. 81). In such cases, even though subjects may lack formalized concepts, there is always an ideal, or target knowledge base against which they can be compared. Content analysis in design, by contrast, is faced with the difficult task of tracking the sorts of elusive references and vague or open-ended concepts that tend to play a role in ill-defined domains.

One study that does, however, attempt to uncover knowledge representations used by designers was done by Akin (78). In one of the experiments reported in that study, the chunking of architectural plans by architects was investigated, following Chase and Simon's (73) analysis of chess board configurations by chess players. Specifically, architects were asked to look at plan drawings and either trace, copy or interpret them, all while verbally reporting their actions. Akin analyzed each subject's behavior by noting pauses in their protocols, ultimately inferring the low-level organization of plan knowledge that they used. In a second experiment, Akin asked architects to try to imagine the contents of a drawing by asking questions about it but without actually seeing it. By analyzing the content and sequence of questions that each subject asked, Akin inferred the structure and content of the inferential knowledge they used.

Another study that focuses on the knowledge used by designers more than on actual design processes is reported in Crismond (97). In that study, a variety of novice and expert designers were given "investigate and redesign" tasks. Initially, subjects were presented with three unlabeled artifacts and asked to identify them. Although the artifacts all performed the same function (opening jars in one series and cracking nuts in another), the subjects knew nothing about them in advance. In addition to being asked what the artifacts did, subjects were also asked a variety of probing questions such as how much they thought the artifacts cost and what type of consumers they were targeted towards. Subjects were subsequently asked to rank the sample artifacts and redesign them. By analyzing verbal reports, Crismond was able to characterize the kinds of device knowledge, scientific knowledge and inquiry skills his subjects possessed. He was also able to track the mapping of knowledge from related domains, finding in one case that experts were more likely to have high-level schemas onto which they could attach and hence integrate individual insights generated from exploratory activities.

Another study, done by Atman et al. (99), focuses on, among other things, activities that help bring additional knowledge to bear on design problems. To generate data, a number of expert and novice engineering students were asked to design a playground. Each subject's behavior was tracked and categorized, but not for the purpose of describing cognitive processes. Instead the frequency of certain types of behavior between the two groups was compared, as was the content of the resulting designs and the strength of correlations between design quality and different types of behavior. One finding resulting from the study is that knowledge seeking activities correlate with design quality for both novices and experts.

A variation on verbal analysis that has been used to study knowledge representations involves analyzing drawings produced by subjects rather than verbal responses. By looking at resulting designs, for example, some researchers have attempted to infer both the kinds of knowledge and the structure of the knowledge used to solve design problems. Similar strategies have been used in studies of creative concept generation in cognitive psychology (e.g. Ward 95), as well as in developmental studies (e.g. Karmiloff-Smith 90). In design, Jansson and Smith (91) use the strategy to study how the presentation of old solutions to a product design problem constrains the generation of new solutions. They found that attributes and structural relations possessed by the old solutions are more likely to be included in new solutions when the old solutions are presented beforehand, even when subjects are told that the old solutions are flawed and should not be copied. They concluded that designers tend to structure their solutions on available exemplars rather than on a priori principles, and seek novelty by varying attribute values rather than structural relations.

4 Process Isolation

Content analysis attempts to overcome the limitations of think-aloud protocols by concentrating on the content of thought, particularly on the kinds of knowledge representations used in design. It does so, however, at the expense of understanding the cognitive processes that mediate representations. A second alternative to think-aloud protocols that aims at cognitive processes as well as at representations, involves the controlled study of processes isolated from drawn-out design episodes. This strategy assumes that a host of different activities undertaken by designers – not just conscious reasoning strategies – may contribute to the transformation of representations in design, leading, for example, to the generation of new problem representations and new ways of structuring knowledge about artifacts. These activities include those traditionally studied in cognitive science, such as analogy (e.g. Gick and Holyoak 83), mental modeling (e.g. Johnson-Laird 83), simulation (e.g. Barsalou 99), conceptual combination (e.g. Wisneiwski 97) and conceptual blending (e.g. Fauconnier and Turner 98).

Research on isolated processes is free to concentrate on fundamental cognitive theories apart from design provided that such theories can ultimately be combined into higher-level accounts of complex problem solving. More often, though, design researchers who focus on isolated processes tend to address the question of how existing theories in cognitive science can be extended to cover design behavior. Some design researchers, for example, have converted experimental tasks commonly used to study isolated cognitive processes into tasks that more closely resemble design or pick up on experimental tasks that already resemble design (e.g. tasks used by gestaltists, like the hat rack design task in Maier 30). The result is often an experimental task that mimics design in some regard but that reduces its complexity to allow for the controlled manipulation of independent variables.

4.1 Process Isolation in Design

Examples of process isolation studies in design include Akin and Akin (96), Casakin and Goldschmidt (99), Lawson (79), and Verstijnen et al. (98). In Akin and Akin (96), processes associated with insight are examined using standard problems from the cognitive science literature (e.g. the nine-dot problem) as well as simple design tasks like the composition of an elevation given a small number of fixed constraints. In Casakin and Goldschmidt (99), analogy is studied by giving designers simple design tasks lasting 10 to 15 minutes while showing them a variety of visual stimuli. In both studies, the tasks given to subjects include controlled miniaturizations of full-blown design problems. In other studies, more abstract tasks, which are taken to be representative of some part of design problem solving, are used instead. In Verstijnen et al. (98), for example, emergence in mental imagery and in drawings is studied using simple patterns and pre-functional shape compositions. Finally, in Lawson (79), a geometrical puzzle is given to both scientists and designers to see how they compare in making inferences about problems and potential solutions.

The trick, in studying isolated processes is coming up with experimental tasks that have well-defined dependent variables that are somehow relevant to design. The more a task is constructed to resemble a real design task, however, the more murky dependent variables may become. In Casakin and Goldschmidt (99), for example, the general quality of resulting designs, as assessed by a panel of experienced designers, is used as the dependent variable. The problem, in this case, is that design quality, although obviously of interest to design researchers, is not a unitary variable. In Casakin and Goldschmidt's study, design quality may be confounded with variables other than those directly related to the analogical processes being studied. And although the finding that design quality is improved by the presence of visual analogs even when subjects are not told to use them is striking, it does not go far to elucidate the underlying cognitive processes. Such studies might be expanded by incorporating content analysis, as was done in the study by Atman et al. (99) discussed earlier, to extract finer-grained dependent variables that are relevant both to design and to the specific processes under investigation.

5 Situated Studies

A final strategy for studying design behavior involves deferring theories of cognitive processes and representations altogether and instead focusing on design activities as they relate to their social, cultural and material contexts – in other words, treating the designer and the social, cultural and material contexts as "intact activity systems" (Greeno 98). In favor of such studies, a variety of researchers in different areas have argued that understanding situated behavior is essential for framing research on lower-level behavior and cognition, much like an understanding of behavior and cognition is essential for framing even lower-level neural theories (Greeno 98; Hutchins 95; Lave and Wenger 91; Suchman 87). These researchers, moreover, call for the development and use of theoretical concepts specific to situated behavior. They argue, in other words, that theories of situated behavior must be developed independently if they are to have a top-down influence on lower-level theories. Working cognitively from the bottom up is seen as insufficient for two reasons. First, low-level processes might interact in unpredictable ways when integrated in more complex behavior, something that is overlooked in the study of isolated low-level processes. Second, low-level research is selective, in that it is framed by common-sense notions of situated behavior; hence, it may overlook processes that, in fact, play a significant role in reasoning. Low-level research framed by more considered investigations of situated behavior will, of course, also be selective, but presumably in a more insightful way.

Studies of situated behavior typically focus on one of two things: the way meaning is produced in situations or the way the social context and material environment regulate behavior. Situated studies focusing on meaning start by pointing out that meaning is not literally carried in external or internal symbols, but rather is constructed in particular situations. Generally, meaning is tied to a symbol's referential status, and it is often through interaction with others and with the environment that one is able to establish (both for the agent and those he or she might be interacting with) the referential context, which includes, at various levels of abstraction, both what is being referred to and the perspective from which it is being referred. On the one hand, this suggests that meaning may be constructed within and for specific situations, arising in a "process of co-construction and negotiation between participants and other systems" (Greeno 98, p. 14). At another level, it suggests that a plan generated in the head (however it happens be externalized) may have a limited significance in studies of behavior since the meaning of a plan ultimately takes shape in a situation (Suchman 87). People, to put it simply, execute plans in action; plans do not execute themselves.

Understanding meaning, of course, is not an end in itself. Meaning ultimately underpins the reasoning practices and continued social and material interactions that are the aim of traditional cognitive studies. This, however, does not mean that situated studies end when the person being studied retreats to a private office to work alone. Problems faced by individuals get their meaning, both as specific problems and as instances of general types of problems, from socio-cultural situations that extend beyond the office, as do the concepts people reason with. Presumably, if one wants to better understand the sorts of concepts an individual is using regardless of their immediate context, studying how they arise through interaction with intact activity systems will yield insights that pure cognitive studies fail to generate.

Situated studies that focus on the way the social context and material environment regulate behavior take a similar approach to studies that focus on the production of meaning. Here, the primary assumption is that social situations and material environments afford certain types of behavior. Just like a cocktail party, for example, affords certain types of behavior, so does a design office or school classroom. In part, individuals become effective practitioners by learning how to participate (Lave and Wenger 91) or, in different terms, how to take advantage of social affordances (Greeno 98). Individuals may also manipulate the environment in order to produce material affordances specific to a given task (e.g.

Kirlik 98). Researchers in this area have focused on the structure of social and material affordances as well as on how individuals interact with and reproduce them. This type of research is seen as particularly crucial if the goal is to improve situations, particularly situations associated with education and professional practice. If, by contrast, one develops an idealized cognitive model and applies it to a specific situation, the social and material contexts may get in the way. The social and material contexts, moreover, cannot simply be removed to make room for idealized practices. Context, rather, is taken to be constitutive of practice. For this reason, one method suggested for studying situated behavior is to actively participate in social and material interventions to see exactly how they interact with existing affordances and how they foster new ones (Greeno 98).

5.1 Situated Studies in Design

Design studies like those discussed in the previous sections are in some respects naturally situated. For one thing, design problems generally take on meaning as they are being worked on – that is, in a situation. A designer may start off with a plan in mind, but plans are generally thought to be of limited value in solving design problems (cf. Ullman et al. 88), especially when compared to well-defined problems. In this sense, design researchers naturally extend the boundaries of traditional cognitive studies when they study design. Additionally, design studies often aspire to some degree of realism. Subjects in protocol studies, for example, are often given a range of realistic resources, sometimes including office environments, artifacts, experts and even teammates. Design problems are also often presented in a realistic fashion. If, for example, a problem used in a protocol must be worked out in an unusually short amount of time, subjects may be told to imagine their task as one of preparing quick concept sketches for a rushed meeting with the client.

Despite the naturally situated nature of some design studies, several factors limit the extent to which experimental settings capture the situated nature of actual practices. First, experimental settings seldom include a persistent social network or a personally tuned physical environment, as one might expect to find in a design office or educational setting. A typical office includes, among other things, coworkers with whom a designer has already developed social relations. Even if a design experiment involves multiple designers working together, in most cases subjects will enter the experimental situation as strangers. In addition, subjects may use concepts in experimental settings that depend on their everyday work environments for meaning. When removed from their everyday settings, these concepts may loose meaning.

Another problem with design studies conducted in experimental settings is that subjects may have difficulty constructing a meaningful image of the client (Lloyd et al. 96) or interactively negotiating meaning with them (although see Dorst 97, for an example of how a subject's view of the client may take shape in a think-aloud protocol). Finally, the social situation in studies of individual designers usually centers on the experimenter and hence offers few affordances for interaction. Although subjects in think-aloud protocols may interact with the experimenter, the experimenter is usually equipped to answer requests for information in a scripted manner, precluding any sort of unstructured communication. In situations outside the lab, however, casual interaction may actually play an important role in reasoning about tentative, speculative ideas (e.g. Roschelle 92; Teasley 94).

While experimental studies are generally lacking in their treatment of situated behavior, there are a handful of studies that do focus specifically on situated aspects of design. Some design studies, both in situ and in the lab, have looked at the role of dynamic interaction with the environment and with other participants. Goel (95), for example, looks at the way interaction with sketches helps facilitate lateral transformations as subjects generate design alternatives. Similar studies dealing with the material environment include Harrison and Minneman (96), which looks at how artifacts in the immediate environment are used improvisationally in team processes to illustrate and reason about alternatives, and Carter (93), which looks at how an architect arranges drawings around his desk to facilitate the retrieval of information within a smooth flow of activities. Carter's observations, as well as those of Minneman and Harrison, and Goel, appear consistent with studies in cognitive science that highlight purposeful manipulations of the environment aimed at making reasoning easier (e.g. Kirsch 95; Hutchins 95). Studies focusing on interaction with other people, on the other hand, include Schon's (82) study of a student interacting back and forth with an expert designer in an actual studio setting. From his observations Schon argues that such interaction helps bring new information into the picture and resolve the designer's understanding of the problem.

A variety of studies in design have also addressed social interaction within design teams, some in actual work settings (e.g. Bucciarelli 94; Lloyd and Deasley 98), some in experimental settings (Cross and Cross 96; Berenton et al. 96). Such studies tend to focus on how the behaviors of different individuals contribute in different ways to a group problem solving process. Some team members might, for example, be observed organizing and collecting votes on design alternatives, while others are observed balancing this sort of behavior by striving to keep design options open. These sorts of studies speculate on group problem solving independent of studies on individual problem solvers. In other words, they tend to treat groups as unique types of cognitive systems. Interaction among team members, for example, may be analyzed in terms of how it overcomes problems like fixation that individuals working outside a group often succumb to (cf. Hutchins 91).

Few studies in design deal specifically with how concepts are constructed, reproduced or conferred through local practices. Most seem to assume that although knowledge may be subjective, it is ultimately derived from experiences via general reasoning (e.g. inductive generalization or pattern recognition). Schon (84), however, discusses the use of design concepts specific to local practices, in particular the role certain exemplars and metaphors play in the architectural program at MIT. He discusses, for example, the nuances and tensions associated with the use of references like "spaghetti bowl" and "Renaissance order." Although he observes such references being used in the architectural studio, he stops short of investigating the broader culture in which they were developed. Thus, one is left with a limited understanding of their content and their place within the institution.

6 Summary

Four distinct strategies for studying design behavior have been discussed in this chapter. Although the strategies were not presented in any particular order, they naturally cluster into two contrasting pairs. Protocol analysis and content analysis, on the one hand, place contrasting emphasis on the transformational moves and representations used by designers. Although they appear more competitive than complementary, both encounter their share of difficulties when applied to design. In particular, while it tends to be difficult to track transformational moves in an ill-defined problem space, it also tends to be difficult to uncover knowledge representations when a subject is drawing on a limitless set of potentially related experiences.

Process isolation and situated studies, on the other hand, take contrasting top-down and bottom-up approaches to studying design behavior. Although they too appear as competing rather than complementary strategies, it is not clear that one outperforms the other. Although process isolation may benefit from the sleuth work of situated studies, process isolation may also inform situated studies. Moreover, process isolation possesses a feature that has been fundamental to the success of science since its inception yet is lacking in the other strategies: the ability to control and measure different phenomena independently.

Not many of the studies discussed in the chapter have, of course, fit neatly into any one of the four strategies covered. Most refer, in fact, to Ericsson and Simon (84) for methodological support but end up using a mix of strategies in pursuit of a mix of goals. Recognizing the multifaceted nature of most research programs, the criticisms laid out in this chapter are intended to be illustrative rather than indicting, probing the strengths and weaknesses of the strategies rather than the individual studies themselves. At this point, however, what is clearly needed is more methodological clarity. Studies may be applauded for developing hybrid research strategies, but not for articulating them. At a minimum, methodological clarity will boost existing research, both by encouraging the continued refinement of research methods and techniques, and by possibly helping researchers merge findings from complementary attacks on design itself.

References

Akin Ö, (1978), How do architects design? *Artificial intelligence and pattern recognition in computer-aided design,* J Latombe Ed. North Holland, New York, NY.

Akin Ö, and Akin C, (1996), Frames of reference in architectural design: Analysing the hyperacclamation (A-h-a-!), *Design Studies*, 17, 341-361.

Akin Ö, and Lin C T, (1996), Design protocol data and novel design decisions, N Cross, H Christiaans, and K Dorst Eds. *Analysing design activity*, John Wiley & Sons, New York, NY, 35-63.

Atman C J, Chimka J R, Bursic K M, and Nachtmann H L, (1999), A comparison of freshman and senior engineering design processes, *Design Studies*, 20(2), 131-152.

Barsalou L W, (1993), Structure, flexibility, and linguistic vagary in concepts: Manifestations of a compositional system of perceptual symbols, A C Collins, S E Gathercole, and M A Conway Eds. *Theories of memory*, Lawrence Erlbaum Associates, London, 29-101.

Barsalou L W, (1999), Perceptual symbol systems, *Behavioral and Brain Sciences*, 22, 577-609.

Berenton M F, Cannon D M, Mabogunje A, and Leifer L J, (1996), Collaboration in design teams: How social interaction shapes the product, N Cross, H Christiaans, and K Dorst Eds. *Analysing design activity*, John Wiley & Sons, New York, NY, 319-341.

Bucciarelli L L, (1994), *Designing Engineers*, MIT Press, Cambridge, MA.

Carter K, (1993), Computer aided design: Back to the drawing board, *Technical Report EPC-1993-107*, Xerox Research Centre Europe.

Casakin H, and Goldschmidt G, (1999), Expertise and the use of visual analogy: Implications for design education, *Design Studies,* 20 (2), 153-175.

Chase W G, and Simon H A, (1973), The mind's eye in chess, W G Chase Ed. *Visual Information Processing*, Academic Press, New York, NY.

Chi M T, Feltovich P, and Glaser R, (1981), Categorization and representation of physics problems by experts and novices, *Cognitive Science,* 5, 121-152.

Chi, M T, (1997), Quantifying qualitative analyses of verbal data: A practical guide, *Journal of the Learning Sciences,* 6(3), 271-315.

Crismond D P, (1997), *Investigate-and-redesign tasks as a context for learning and doing science and technology: A study of naïve, novice and expert high school and adult designers doing product comparisons and redesign tasks*, Doctoral dissertation, Harvard University.

Cross N, and Cross A C, (1996), Observation of teamwork and social processes in design, N Cross, H Christiaans, and K Dorst Eds. *Analysing design activity*, John Wiley & Sons, New York, NY 291-317.

Dorst K, (1997), *Describing design: A comparison of paradigms*, Doctoral dissertation, Technical University of Delft.

Dorst K, and Dijkhuis J, (1996), Comparing paradigms for describing design activity, N Cross, H Christiaans, and K Dorst Eds. *Analysing design Activity*, John Wiley & Sons, New York, NY, 253-269.

Eastman C M, (1969), Cognitive processes and ill-defined problems: a case study from design, *Proc. of the first joint international conference on artificial intelligence,* Washington, DC, 675-699.

Ericsson K A, Simon H A, (1993), *Protocol analysis: Verbal reports as data*, MIT Press, Cambridge, MA.

Fauconnier G, and Turner M, (1998), Conceptual integration networks, *Cognitive Science,* 22 (2), 133-187.

Gero J S, and McNeill T, (1998), An approach to the analysis of design protocols, *Design Studies*, 19(1), 21-61.

Gero J S, (1990), Design prototypes: A knowledge representation schema for design, *AI Magazine,* 2(4), 26-36.

Gick M, and Holyoak K, (1983), Schema induction and analogical transfer, *Cognitive Psychology,* 15(1), 1-38.

Goel V, (1995), *Sketches of thought*, MIT Press, Cambridge, MA.

Goldschmidt G, (1996), The designer as a team of one, N Cross, H Christiaans, and K Dorst Eds. *Analysing design activity*, John Wiley & Sons, New York, NY, 65-91.

Greeno J, (1998), The situativity of knowing, learning, and research, *American Psychologist*, 53, 5-26.

Harrison S, Minneman S, (1996), A bike in hand, N Cross, H Christiaans, and K Dorst Eds. *Analysing design activity*, John Wiley & Sons, New York, NY, 417-436.

Hutchins E, (1991), The social organization of distributed cognition, L B Resnick, J M Levine, and S D Teasley Eds. *Perspectives on socially shared cognition*, American Psychological Association, Washington, DC.

Hutchins E, (1995), *Cognition in the wild,* MIT Press, Cambridge, MA.

Jansson D G, and Smith S M, (1991), Design fixation, *Design Studies*, 12(1), 3-11.

Johnson-Laird P N, (1983), *Mental models: Towards a cognitive science of language, inference, and consciousness*, Harvard University Press, Cambridge, MA.

Karmiloff-Smith A (1990), Constraints on representational change: Evidence from Children's Drawing, *Cognition*, 34, 57-83.

Kirlik A, (1998), The ecological expert: Acting to create information to guide action, *Proceedings of the fourth symposium on human interaction with complex systems*, Dayton, OH.

Kirsh D, (1995), Complementary strategies: Why we use our hands when we think, *Proc. of the seventeenth annual conference of the cognitive science society*, Lawrence Erlbaum and Associates, Hillsdale, NJ.

Lawson B R, (1979), Cognitive strategies in architectural design, *Ergonomics,* 22(1), 59-68.

Lloyd P, and Deasley P, (1998), Ethnographic description of design networks, *Automation in Construction*, 7, 101-110.

Lloyd P A, and Scott P J, (1994), Discovering the design problem, *Design Studies*, 15(2), 125-140.

Lloyd P A, Lawson B R, and Scott P J, (1996), Can concurrent verbalisation reveal design cognition? N Cross, H Christiaans, and K Dorst Eds. *Analysing design activity*, John Wiley & Sons, New York, NY 437-462.

Loftus G R, and Bell S M, (1975), Two types of information in picture memory, *Journal of Experimental Psychology: Human Learning & Memory,* 1(2), 103-113.

Maier N R, (1930), Reasoning in humans, I: On direction, *Journal of Comparative Psychology*, 10, 115-143.

Minsky M, (1975), A framework for representing knowledge, P H Winston Ed. *The psychology of computer vision,* McGraw-Hill, New York, NY.

Newell A, and Simon H A, (1972), *Human problem solving,* Prentice-Hall, Englewood Cliffs, NJ.

Nisbett R, and Wilson T, (1977), Telling more than we can know: Verbal reports on mental processes, *Psychological Review*, 84(3), 231-259.

Nisbett R E, and Ross L, (1980), *Human inference: Strategies and shortcomings of social judgment*, Prentice-Hall, Englewood Cliffs, NJ.

Roschelle J, (1992), Learning by collaborating: Convergent conceptual change, *Journal of the Learning Sciences*, 2, 235-276.

Rummelhart D, and Ortony A, (1977), The representation of knowledge in memory, R C Anderson, R J Spiro, and W E Montegue Eds. *Schooling and the Acqusition of Knowledge*, Lawrence Erlbaum and Associates, Hillsdale, NJ.

Schank R C, and Abelson R P, (1977), *Scripts, plans, goals and understanding: An inquiry into human knowledge structures,* Lawrence Erlbaum and Associates, New York, NY.

Schon D A, (1982), *The Reflective practitioner: How professionals think in action*, Basic Books, New York, NY.

Schon D A, (1984), Problems, frames and perspectives on designing, *Design Studies*, 132-136.

Schooler J W, and Engstler-Schooler T Y, (1990), Verbal overshadowing of visual memories: Some things are better left unsaid, *Cognitive Psychology,* 22(1), 36-71.

Schooler J W, and Melcher J, (1995), The ineffability of insight, S Smith, T Ward, and R Finke Eds. *The creative cognition approach*, MIT Press, Cambridge, MA.

Simon H A, (1973), The Structure of Ill Structured Problems, *Artificial Intelligence,* 4, 181-201.

Teasley S D, (1997), Talking about reasoning: how important is the peer in peer collaboration? L B Resnick, R Saljo, C Pontecorvo and B Burge Eds. *Discourse, tool and reasoning: Essays on situated cognition*, Springer, New York, NY, 361-384.

Ullman D G, Dietterich T G, Stauffer L, (1988), A model of the mechanical design process based on empirical data, *Artificial Intelligence in Engineering Design and Manufacturing*, 2 (1), 33-52.

Verstijnen I M, Hennessey J M, Leeuwen C van, Hamel R, and Goldschmidt G, (1998), Sketching and creative discovery, *Design Studies*, 19(4), 519-546.

Ward T, (1995), What's old about new ideas, S M Smith, T B Ward, and R A Finke Eds. *The Creative cognition approach*, MIT Press, Cambridge, MA.

Wisniewski E J, (1997), Conceptual combination: Possibilities and esthetics, T B Ward, S M Smith, and J Vaid Eds. *Creative thought: An investigation of conceptual structures and processes*, American Psychological Association, Washington, DC.

Chapter 3

Studying Engineering Design Learning: Four Verbal Protocol Studies

Cynthia J. Atman and Jennifer Turns
Center for Engineering Learning and Teaching
Industrial Engineering
University of Washington
Seattle, WA

Abstract

Design is a hallmark of engineering. In the last decade, engineering design has become a central topic in discussions of engineering education reform, leading to a strong need to understand the processes of engineering student designers. In this chapter, we focus specifically on the use of verbal protocol analysis (VPA) to characterize the design learning and knowing of students in engineering schools. Over the past six years, we have conducted a series of four VPA studies of engineering students' approaches to design problems. In this chapter, we describe how we applied the VPA methodology to the study of engineering student design processes. We also summarize the results of each of the four studies. In the discussion, we start a synthesis of the overall results and present some ideas about the implications of the results for engineering design education.

1 Engineering Design Learning

Engineers must know how to design to be effective practitioners. Engineers design or redesign a variety of products and processes such as biomaterials for surgical implants, database tools, quality control equipment, procedures for handling hazardous materials, and supply chain distribution processes.

Although design has always been recognized as a central engineering activity and an important issue in engineering education (Mann 18), there has been a growing awareness in recent years of the need to emphasize design learning in engineering curricula and the need to learn how to teach design better (NRC 95; NSF 95). This has been reflected in the revised accreditation standard for engineering education where "an ability to design a system, component, or process to meet desired needs" is one of the eleven outcomes engineering students are expected to achieve by graduation (ABET 00). Increasingly, the ubiquitous capstone design experience of engineering schools are being supplemented by freshman design experiences as well as design experiences at other points in engineering curricula (Todd et al. 95; Sheppard and Jenison 97). These observations and issues all suggest the time is right for research that can inform the teaching of engineering design.

Coincidentally, the past two decades have seen a surge in the amount of research available on engineering design. Some researchers have focused on exploring normative or prescriptive models to guide engineering design activity (Suh 90; Pahl and Beitz 96). Others have focused on understanding actual design behavior, using a range of research approaches. For example, Bucciarelli (94) used ethnographic approaches to study design behavior of practicing engineers over the several months required to complete professional design problems. Radcliffe and Lee (1989) used a form of content analysis to study the design activities of freshman engineers in their week-long response to a problem statement. Tang and Leifer (91) have been studying the design performance of engineers using a video-based methodology and one to two-hour design problems. Cross and his colleagues have compiled multiple analyses of video-taped protocols of team design processes into a single book (Cross et al. 96). In general, entire journals such as *Research in Engineering Design* explicitly provide insight into engineering design, while other design journals such as *Design Studies* regularly discuss engineering design issues.

Collectively, such studies have resulted in a rich set of information available to inform engineering design education. For example, study results point to goal states for engineering design education (e.g. processes students may aspire to), acceptable variability in successful processes, and strategies that may influence the success of a design process. However, the application of such insights to engineering design education can be difficult without better understanding the starting point and the development of students as they are learning about engineering design. If we take the famous Ausebel quote to heart -- "If I had to reduce all of educational psychology to just one principle, I would say this: The single most important factor influencing learning is what the learner already knows. Ascertain this and teach him accordingly" (Ausebel 68) -- then it seems that gaining an understanding of engineering student design processes needs to be a central element of research on engineering design that can inform engineering design education.

It has been our goal to characterize what learners in engineering schools know about design. To work toward this goal, we have been studying the processes of engineering students approaching design problems and characterizing these processes in a way that can lead to pedagogical implications. Our means for gaining detailed insight into the processes of engineering design has been the use of verbal protocol analysis.

In the period since 1993, we have conducted four studies of engineering student design behavior using the verbal protocol analysis research methodology. The goal of this chapter is to survey these four studies and suggest some implications of the results. Taken together, the results of these studies shed light on the starting and ending points of engineering students as they learn about engineering design and learn to solve engineering design problems. The next section starts by explaining the verbal protocol analysis methodology and our application of verbal protocol analysis in the four studies. The four studies, which collectively explore the design processes of freshmen and seniors on four different design problems, are summarized in the next section. The discussion focuses on synthesizing the results of the four studies and identifying possible implications of the studies for engineering design education.

2 Verbal Protocol Analysis: A Method for Studying Engineering Student Design Activity

Verbal protocol analysis is a rigorous methodology used to document and understand the processes associated with cognitive activities, such as design (Ericsson and Simon 93). In a verbal protocol study, subjects complete experimental tasks while concurrently talking out loud. The verbal protocol then becomes the data to be analyzed in order to give insight into the cognitive process of interest. The assumption behind verbal protocol analysis is that as long as the subject is simply talking about what they are doing, the talking will not significantly change the approach they use to perform the activity (Ericsson and Simon 93; Oxman 95). Verbal protocol analysis has been used in cognitive science to study and understand a great variety of cognitive behaviors from problem solving (e.g. Schoenfeld 85), to learning strategies (e.g. Ng and Bereiter 91), to design (e.g. Ullman et al. 88; Guindon 90; Ennis and Gyeszley 91; Christaans and Dorst 92; Rowland 92; Sutcliffe and Maiden 92; Cross et al. 94; James et al. 94; Cross et al. 96).

The implementation of verbal protocol analysis can be time-consuming and the results challenging to interpret. However, the process of verbal protocol analysis research is relative straightforward (e.g. Chi 97). The process consists of the following general activities (not necessarily in this order):

1. Develop a coding scheme,
2. Choose a problem, in this case a design problem,
3. Collect protocols from students as they solve the problem,
4. Code the protocols according to the coding scheme, and
5. Analyze and interpret results.

Our application of these activities in each of the four studies has been almost identical. In what follows, we describe our coding schemes, problems, data collection procedure, coding procedure, and analysis and interpretation approach. What we present in this section represents our current approach to using verbal protocol analysis and the approach used in the latter studies. More extensive insight into our application of the verbal protocol analysis can be found in other publications (Atman et al. 96; Atman and Bursic 98). Because we have been doing this work for several years and because our application of the method has evolved, the use of verbal protocol analysis in the earliest studies differs slightly from the description presented below. In the next section, when we describe the details of specific studies, we will explain the variations in method.

Develop Coding Scheme. In a verbal protocol study, the coding scheme is the link between the verbal protocols and the hypotheses being tested and/or the questions being asked about the cognitive processes of interest.

Table 1. The Coding Scheme

Design Activity	Description
Identification of Need	Identify basic needs (purpose, reason for design)
Problem Definition	Define what the problem really is, identify constraints and criteria, reread or question the problem statement
Gathering Information	Search for and recall information beyond that provided.
Generating Ideas	Develop possible solution ideas, list different alternatives.
Modeling	Describe how to build the idea, how to make it, measurements, dimensions, calculations
Feasibility Analysis	Determine workability, verification of workability, does it meet constraints, criteria, does it make sense.
Evaluation	Compare alternatives (is one better, cheaper), make judgments about one alternative relative to another.
Decision	Select one idea or solution among alternatives
Communication	Define the design solution to others, labeling the final design, or writing down instructions.
Implementation	Produce or construct a physical device, product, or system

Since our central interest has been in engineering design processes, our primary coding scheme has embodied a set of activities associated with engineering design. Our current coding scheme is presented in Table 1. The coding scheme identifies ten "elemental" activities associated with design including identification of need, problem definition, information gathering, generation of alternatives, modeling of alternatives, determining feasibility of alternatives, evaluating alternatives relative to each other, deciding among alternatives, and communicating the results of the design process to someone else.

Essentially, the coding scheme represents a set of activities in designing a solution to an open-ended problem.

This design coding scheme is based on analysis of the design processes presented in several engineering textbooks (Moore et al. 95). However, the coding scheme does not imply a specific sequence or duration of the activities. As we have seen repeatedly in our research, different *sequences* and *durations* of these activities can lead to a variety of different *design processes.*

Additionally, we have developed coding schemes to describe other aspects of student design behavior. For example, we have explored the number and type of criteria that students considered. With a similar approach, we have explored the number of design alternatives that students consider. We have also developed coding schemes to explore information gathering behavior. In this case, surveys of user populations and literature searches were used to develop a master set of categories of information relevant to the problem. The protocol was then coded using these categories. Such an approach permitted us to characterize when and how subjects gathered information and to explore the range of information that was collected.

Choose a problem. In verbal protocol research, the coding scheme is used to characterize behavior exhibited as subjects engage in some type of cognitive activity. The behavior, in turn, is typically elicited by having subject's complete some type of problem. The choice of those problems in verbal protocol studies is typically constrained by the requirement that subjects "think-aloud" during the process of completing the problem and also by the need to record and analyze the protocol.

Collectively, our studies have involved the use of four design problems. In three of these problems, subjects are asked to design a solution to satisfy a stated need. In the other problem, subjects are asked to describe factors that would need to be taken into account during the design process. In all of the problems, effort was made to select problems that can be tackled as easily by freshmen as by seniors. The wording of each of these problems is given in Table 2. Some background on each problem is given below:

- The Ping-Pong problem. In this problem, the students are asked to design a ping-pong ball launcher. This problem was chosen to resemble a textbook problem (ECSEL 93). In our experiments, thirty minutes are typically allocated for subjects to solve the problem.
- The Street Crossing problem. In this problem, the students are asked to design a means for people to cross a busy street. In particular, the "busy street" referred to in the problem statement was a street that the subjects in all experiments using this problem crossed on a daily basis. The problem was developed in order to explore whether students design differently when they can rely on personal experiences such as the

daily crossing of the street. In our experiments, thirty minutes are typically allocated for subjects to solve the problem.

- The Playground problem: In this problem, students are asked to design a playground for a fictitious neighborhood. The problem was developed in order to explore student design behavior in designing something with which all students are familiar. In order to explore information-gathering behavior, the experimental protocol for this problem included subjects being able to ask the experimenter for information. In our experiments, three hours have been allocated for subjects to solve the problem.
- Midwest Floods: In this problem, subjects are asked to identify factors to be considered in the design of a retaining wall for the Mississippi river. This problem was developed as a way to explore the broadness with which students define a design problem – issues in the design of a retaining wall range from purely technical considerations to large social and environmental considerations. In our experiments, thirty minutes have typically been allocated for subjects to complete the problem (Table 2).

Since each of the first three problems asked subjects to design a solution, part of the design of the problem was the development of a strategy to determine solution quality. Specifically, coding schemes were developed to determine the quality of each of the design solutions. Each of these schemes consists of (1) a set of dimensions that would be addressed in the ideal solution and (2) point values to be included in the quality score if the design addresses the dimension. These dimensions were determined from discussions with experts and a review of relevant literature. The quality scores for subjects' designs were determined by having two coders independently score the quality based on the dimensions, determine the reliability of their coding, and arbitrate the disagreements to consensus.

Have subjects solve problem. The heart of a verbal protocol analysis study is the point where the subjects solve the chosen problem while concurrently providing verbal protocols associated with their actions. Since the subject is audio-taped or video-taped during the process, verbal protocol analysis puts constraints on the environment in which the activity can be carried out.

Table 2. Text from the problems used in the studies

Ping Pong Launcher. In an attempt to avoid boredom at Benedum Hall, creative engineering students developed a challenging new game. A ping pong ball is to be launched at a bullseye target, and points are awarded according to the accuracy of the landing. However, the ping-pong ball cannot be thrown at the target. It is up to you to design a device which will lift the ping-pong ball into the air and land it at the target. An accurate landing is desired while also maintaining a long flight time. Given that the entire launching assembly must not be greater than 1m x 1m x 1m in dimension, design a ping-pong ball launcher for this game. Your work should contain a detailed description of your design and should include any relevant diagrams and calculations. Please clearly state all assumptions which are needed in your analysis and try to keep your design simple yet effective.

Street Crossing. College campuses are often overcrowded with pedestrians crossing the streets, since walking is a popular form of transportation for college students. One busy intersection at Pitt is the crossing of Fifth avenue in front of the bookstore. Dangers at this intersection include heavy traffic and busses which run against the general traffic flow. The University would like to design a cost effective method to cross Fifth Avenue which would reduce the possibility of accidents at this intersection. Your work should contain a detailed description of your design and should include any relevant diagrams and calculations. Estimate both the costs and the benefits association with your design. Please clearly state all assumptions which are needed in your analysis and try to keep your design simply yet effective.

Midwest Floods. Over the summer, the Midwest experienced massive flooding of the Mississippi River. What factors would you take into account in designing a retaining wall system for the Mississippi?

Playground. You live in a mid-size city. A local resident has recently donated a corner lot for a playground. Since you are an engineer who lives in the neighborhood, you have been asked by the city to design a playground.

You estimate that most of the children who will use the playground will range from 1 to 10 years of age. Twelve children should be kept busy at any one time. There should be at least three different types of activities for the children. Any equipment you design must be safe for the children, remain outside all year long, not cost too much, and comply with the Americans with Disabilities Act.

The neighborhood does not have the time or money to buy ready made pieces of equipment. Your design should use materials that are available at any hardware or lumber store. The playground must be ready for use in 2 months.

Please explain your solution as clearly and completely as possible. Someone should be able to build the playground from your solution without any questions. The administrator has a lot more information to help you address this problem if you need it. Be as specific as possible in your requests.

For example, if you would like a diagram of the corner lot, some information about the lot appearance, etc., you may ask for it now. If you think of any more information you need as you solve the problem, please ask for it.

Remember, you will have approximately three hours to develop a complete solution. The administrator will tell you how much time is left while you work.

In all of our studies, subjects solved the problem(s) individually in a quiet laboratory environment. Subjects had one large desk space (in sight of the video camera) and a supply of paper and pencils. During the recruiting process, subjects were given an expectation of how long the problems might take, based on anticipated upper bounds on the amount of time that subjects might need. Within this constraint, subjects were able to use the amount of time they needed. In three of the four studies, the subjects solved the problem in the absence of access to information resources such as reference books. In the other problem (the playground problem), the subjects could ask for information from the experimenter. In all cases, the goal was generally to determine what the students could do as designers, with a minimum of resources available.

In each case, the experimental session opened with the subject being introduced to the experiment. The subject then worked a practice think-aloud protocol problem, as suggested by Ericsson and Simon (Ericsson and Simon 93). The bulk of each experimental session was taken up by the subject's efforts to solve the design problems. Throughout the session, the subject provided think aloud protocols, commenting on their activities. If the subject became quiet for a long time, the experimenter would remind the subject to provide the protocols. Finally, subjects were asked to supply demographic data.

The subjects across the four studies included freshmen who had not declared a major and seniors from different engineering majors. All subjects were recruited from the engineering student population at a single university. All subjects were paid for their participation.

Code the protocols. The verbal utterances resulting from the think-aloud process represent the data in a verbal protocol analysis study. The goal of the coding is to map the coding scheme onto the protocol in order to address the research questions. This can be done using a variety of strategies (Chi 97). In general, the challenge is to achieve a high degree of replicability.

Our approach to coding the protocols has been guided by a desire to achieve the greatest degree of replicability and rigor possible. As a result, we have used the following process in each of our four studies. We have transcribed the entire protocol (either professionally or with the help of undergraduate researchers), segmented the transcribed protocol into idea units, and applied the coding scheme to these idea units.

To specifically address replicability, all segmenting and coding have been done by two independent coders. For the coding to be acceptable, the initial reliability of the segmenting or coding has had to meet a minimum standard of 80%. Assuming the reliability level is met, the coders then negotiate to consensus any disagreements. If the desired reliability is not met, the work is redone by a different set of coders. Because this process is person intensive, undergraduate student researchers trained in the segmenting and coding processes have completed much of the segmenting and coding work.

Analyze and Interpret Results. Upon the completion of the coding, it becomes possible to characterize the activity in the protocol through the theoretical lens captured by the codes. Two main ways to depict the coding results are the use of summary metrics and the use of graphical representations of individual activity.

Across our analyses, we have reported a variety of summary metrics. Some standard metrics have been used across all of the studies, including extent of effort devoted to the design process and the number of transitions among design steps. Others, such as measures of the amount of information gathered, have been used in only a subset of the studies. In the presentation of the results of the studies, we clarify the set of metrics used in each study.

While these summary statistics do permit us to understand the trends between groups of subjects, they do not permit us to truly understand the details of individual design processes. For this, we have identified a variety of graphical representations that are useful for better understanding the data. For example, the timeline representation shows the sequence of activity over time. The example timelines in Figure 1 (included later in the paper in the context of one of the studies) represent the activities of freshmen and seniors who designed playgrounds of differing quality. In the timelines, the x-axis represents time during the experimental session and the y-axis corresponds to design activities used in the coding scheme. The tick marks in the timeline represent instances of an activity. The location and width of the tick mark indicate the starting time and duration of the activity, respectively. Although the timelines have been our most influential graphical representation, we have explored and learned from other representations, as described in our papers elsewhere (Chimka and Atman 98).

3 VPA Studies of Engineering Student Design Activity

In this section, we present the four studies we have conducted using the Verbal Protocol Analysis methodology described in the previous section. Table 3 summarizes the four studies.

3.1 Study 1 - Freshmen Learning After One Semester

The first study conducted using verbal protocol analysis was the Freshman Semester Study (Mullins and Atman 94; Mullins et al. 99). The goal of this study was to determine the impact of the first semester of engineering education on students' design processes. An underlying goal was to illustrate the benefits of the verbal protocol analysis methodology as a means for characterizing engineering design processes.

Table 3: The Studies

STUDY	Subject Pool			
	Freshmen	Sophomores	Juniors	Seniors
Freshman Semester Study (Mullins et al., 1999) - Pre and Post 1st-Semester Freshman across subject study - Ping Pong, Street Crossing, Midwest Floods - Data Collected: 1993	| |[1] 16 16			
Freshman Design Text Study (Atman and Bursic, 1996) - Text reading intervention across subjects - Problems Used: Ping Pong, Street Crossing - Data Collected: 1994	| 5/5 [2]			
Freshman/Senior Comparison Study (Atman et al., 1999) - Design performance of freshman and seniors, across subjects - Problems used: Playground Problem - Data Collected: 1994-1995	| 26			| 24
Senior Follow-Up Study - Design performance of seniors (some from freshman study) - Problems Used: Ping Pong, Street Crossing, Midwest Floods - Data Collected: 1997-1998	| |[3] 16 16			61/ 18[4]

[1] The tick-marks indicated the period, during the academic year, when the protocols were collected. The numbers below the tick-marks indicate the number of protocols that were collected.
[2] In the design text study, five subjects solved the design problems without reading the text while the other five read the text first. A total of ten subjects participated in the study.
[3] In the senior follow-up study, the protocols originally collected in the freshman semester study will be re-coded and compared with new protocols. The pale lines indicate the carry over of the subjects.
[4] In the senior follow-up study, 18 of the 61 senior protocols were collected from subjects who had also participated as freshman.

To explore the impact of a semester of freshman engineering education, this study was set up to compare the design behaviors of freshmen at the beginning of their first semester (pre-engineering) with the design behaviors of freshmen at the end of their first semester (post-semester). These subjects were asked to work through the Ping-pong, Street Crossing, and Midwest floods problems in a session scheduled for two hours. The study was set up as a between-subject design, with 16 pre-semester subjects and 16 post-semester subjects. The transcribed protocols were coded using a design coding scheme that is a subset of our current coding scheme. In particular, feasibility analysis was not included as a separate step and the information gathering code was not used.

Data from the three problems show that the students that had taken a semester of the engineering curriculum 1) exerted more effort in solving the problems, 2) had more transitions among the steps in the design process, and 3) considered more criteria in their design process than did beginning students. However, there were no significant differences in design quality.

In this paragraph, we report averages and statistical test results for specific metrics used in the analysis. Level of effort was measured via the number of words spoken by subjects in the protocol. The number of spoken words was significantly greater for post-semester subjects for both the Ping Pong problem (596 vs. 239, $p=0.002$) and the Street Crossing problem (433 vs. 199, $p=0.007$). Also, the number of iterations was significantly greater for post-semester subjects across both the Ping Pong problem (6.49 vs. 2.69, $p = 0.011$) and the Street Crossing problem (5.94 vs. 3.63, $p = 0.002$). In the analysis of the Midwest floods data, it was learned that post-semester seniors identified significant more criteria (10.13 vs. 6.60, $p=0.03$). These results suggest that engineering students are already becoming better designers after their first semester.

3.2 Study 2 - Learning From a Design Text

A challenge of interpreting the results of the first study lies in knowing what part of the intervention (the semester of university engineering education) is responsible for the change in design processes. While elements of the formal education (e.g., classes and laboratories) could be responsible for the changes in design process, it is also possible that other aspects of the semester experience (e.g., living away from home) could have contributed to the observed changes.

In light of this, a goal of the second experiment was to determine if a small controlled intervention would have measurable (and similar) effects on design processes (Atman and Bursic 96). The intervention selected was one of the smallest meaningful interventions that could be identified – having students read part of a textbook chapter on engineering design process. Thus, the second study explored the impact on student design processes of reading such a text.

In the second study, one half of the subjects (first-year engineering students) read a design text before solving three short open-ended engineering design problem. The other half read the text during the experiment, but after the problem solving session. One half of the subjects did not. As indicated in Table 3, the study included ten subjects – five subjects in each condition. The subjects solved the same three problems used in the previous study – the Ping Pong, Street Crossing, and Midwest Floods problems.

Despite the smaller number of subjects and the much smaller scope of the intervention, the results of this study were quite similar to the results in the first study. For the Ping Pong and Street Crossing problems, there were no significant differences in the quality of the solutions but several differences in design process. Subjects exerted significantly more effort (as measured by words spoken, Ping Pong $p = 0.0015$, Street Crossing $p = 0.008$). The subjects also had significantly more transitions among design steps (Ping Pong $p=0.05$, Street Crossing, $p=0.003$). Although the results were not statistically significant, there were trends that subjects solving each of these problems identified more alternative solutions and included activity associated with more of the design steps. For the Midwest Floods problem, subjects who read the design text before solving the problems identified significantly more design factors ($p=0.016$).

The design timelines in Figure 1 show the subjects' design processes for solving the Ping Pong problem. On the left of the figure are the timelines for subjects who did not read the text before solving the problem. On the right are the timelines for subjects who did read the text before solving the problem. In these timelines, unlike our more recent timelines, the marks on the timeline correspond to one "idea unit" from the transcript rather than being proportional to time.

The timelines clearly demonstrate the finding that subjects who read the text before solving the problem had a greater number of transitions among design steps. The timelines also show the greater effort exerted by the subjects who read the design text, and how that effort was allocated across design activities. The timelines also demonstrate the great variability in design processes of the subjects. For example, the last subject in each group seems to have a linear model that involves little cycling back to previously visited steps. In contrast, several other subjects have processes that involve them cycling back to previous visited design steps, such as problem definition. This behavior is very noticeable in the top two timelines of subjects who read the design text, where both subjects return frequently to problem definition types of activity.

3.3 Study 3 - Comparing Freshmen and Seniors

The first two studies explored the design processes of freshmen solving relatively short design problems. The studies show that engineering students are already becoming better designers after their first semester and that even small interventions such as reading a textbook can affect the design process. These studies also succeeded in demonstrating the viability of verbal protocol analysis as a means for studying the design processes of engineering students.

These studies paved the way for questions in several new areas -- the design processes exhibited on longer problems, the design processes exhibited when the designer can gather information, and differences in design processes among freshmen and seniors. The third study sought to expand the scope of the existing research by addressing questions in each of these areas (Atman et al. 95; Bursic and Atman 97; Chimka et al. 97; Adams and Atman 99; Adams and Atman 99; Atman et al. 99; Umphred et al. in-progress).

In the third study, subjects were asked to design a playground for a fictitious neighborhood (see Table 2). Because students were assumed to have had experience with playgrounds, it was expected that the students would engage in this problem for longer than they had engaged in the previous problems. To enable a rich yet controlled exploration of information gathering behavior, the development of the experimental protocol included an effort to identify the types of information that might useful in designing a playground so that the information could be made available to the subjects. During the experiment, subjects were encouraged to ask the experimenter for specific types of information that they would like to have. The log of these requests became part of the data in the study. As indicated in Table 3, twenty-four seniors and twenty-six freshmen participated in the study.

In the analysis, the data were coded in a number of different ways. The design step codes were used as before to capture design process. The data were also coded for the type of information used by the subject and the different playground objects mentioned (e.g., swingset, parkbench). In addition, a new design process variable – design stage – was introduced. The three codes of this variable – problem scoping, developing alternatives, and project realization – were derived from the design step codes in the following way – problem scoping consists of identification of need, problem definition, and information gathering; developing alternatives consists of generating alternatives, modeling, feasibility analysis, and evaluation; and problem realization consists of decision making, communication, and implementation.

One of the most basic results to emerge from this study was that seniors achieved a significantly higher quality score than freshmen ($p=0.046$). The quality score was made of up of three equally weighted ratings concerning how well the subject addressed (1) general playground design considerations, (2) specific equipment design considerations, and (3) general design considerations. On the scale ranging from 0 to 1, freshmen averaged 0.45 and seniors averaged 0.51. Such quality differences had not been identified in the previous studies.

In terms of process differences, the results of this study confirmed many, but not all of the findings in previous studies. To start with, unlike previous studies, there was not a significant difference in the average effort exerted across the two groups of subjects. Freshmen and seniors did not spend significantly different amounts of time on the problem.

However, what seniors did with their time was found to be quite different (Chimka et al. 97; Atman et al. 99). Here we present averages and statistical test results for several measures used in the analysis. Concerning the overall process, seniors had significantly more transitions between design steps (seniors: 186.4, freshmen: 114.9, p=0.002). Seniors also had a greater rate of transitioning among design steps (seniors: 1.8/min, freshmen: 1.2/min, p=0.025).

Concerning the distribution of behavior across different design activities, seniors were found to progress further into the design process than freshmen. For example, seniors spent significantly more time in the decision step than freshmen (seniors: 44.9 seconds, freshmen: 24.3 seconds, p=0.04). Seniors also spent significantly more time in the project realization phase. Seniors spent an average of 5.6 minutes in project realization while freshmen spent an average of 3.0 minutes (p=0.016). While one might imagine that seniors would also have spent a greater allocation of their effort engaged in problem definition, this was not found to be the case.

Seniors also outperformed freshmen on several information gathering measures (Atman et al. 95; Bursic and Atman 97). Seniors had more explicit information requests – requests to the experimenter for information (seniors: 25, freshmen: 14.2, p=0.015). Seniors also made more assumptions than freshmen (seniors: 9.3, freshmen 2.8: p=0.0001). In their overall information gathering performance, seniors requested information associated with a greater number of the information categories (seniors: 6.8, freshmen: 4.1, p=0.006). Given that sixteen categories of information had been identified, these numbers suggest that both groups failed to consider a great deal of the information that would have been important. For example, as a whole, students did not adequately consider factors such as what children like to play with, access to bathroom facilities, accessibility for handicapped children, legal liability, and facility maintenance.

Many of these findings are illustrated in the timelines of freshmen and senior subjects shown in Figure 1 (published previously in (Atman et al. 99)). The top of Figure 1 contains timelines for three freshmen, two receiving a low quality score (Freshman One and Freshman Two) and one receiving a high quality score (Freshman Three). The botton of Figure 1 contains timelines for two seniors, one receiving a low quality score (Senior One)

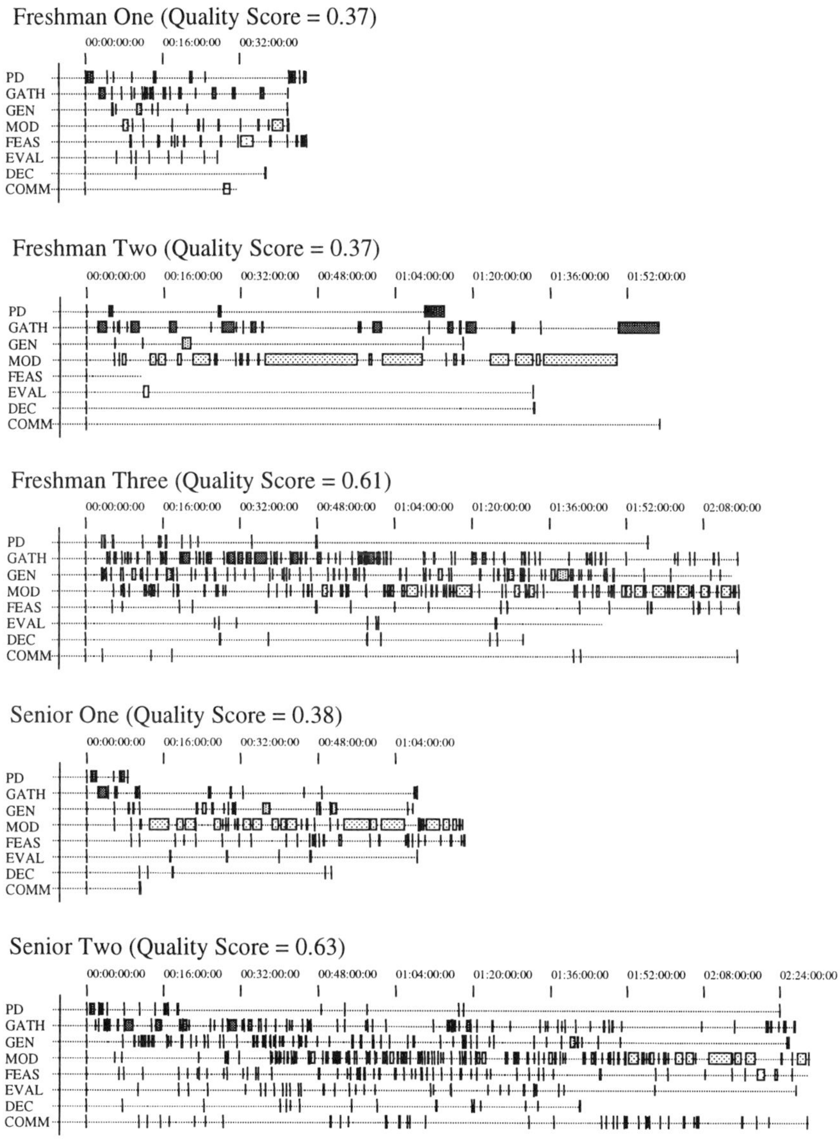

Figure 1: Timelines from Freshmen and Seniors in the Playground Study

and one receiving a high quality score (Senior Two). The width and distribution of tickmarks across all of the timelines provide a vivid visualization of the transition results. The timeline of Freshman Two illustrates a commonly expected novice design behavior – pick a single alternative and spend a great deal of energy developing a the solution via modeling. Although both senior subjects spent a large portion of their time in modeling, as did Freshman Two, neither senior subject had the same types of uninterrupted modeling that Freshman Two did. The timeline of Freshman One illustrates a novice behavior that was discovered through the analysis – getting stuck in problem scoping. This behavior is the complement of the senior behavior of progression to the product realization stage.

Ongoing work with this data is proceeding in a number of directions. One effort is focused on the development of models relating quality to specific design process variables (Umphred et al. in-progress). Preliminary findings using regression modeling suggest that the models for freshmen and seniors are different. The best model to explain freshman quality scores seems to include three variables: the number of explicit information requests, the number of playground objects mentioned, and the time spent in modeling. In contrast, the best model for seniors seems to include only two variables: the number of transitions among design steps and the number of playground objects mentioned.

Two other efforts have been focusing on different ways to characterize process. One area of inquiry has been to look at the design process in terms of how the subject is talking about the design product rather than what they are engaged in doing with their time (Turns et al. 99). Following work by Gero and others (Gero and McNeill 99), we have been looking at how subjects move between functional, behavioral, and structural descriptions of their design. Another effort, work by Adams, has been focusing on identifying and characterizing iteration behavior (Adams and Atman 99; Adams and Atman 99). Iteration is one aspect of design behavior that is almost universally considered important. This current research puts such ideas to the test by attempting to operationalize iteration and characterize differences in the patterns of iteration of freshmen and seniors.

3.4 Study 4 - Senior Follow-up to Freshman Semester Study

The fourth study (currently under analysis) complements the previous three studies in several ways. Like the first two studies, the current study uses the Ping Pong, Street Crossing, and Midwest Floods problems. Like the third study, the focus is on differences between freshmen and seniors. The fourth study is unique because a portion of the data is within-subject data.

The fourth study has been entitled the "senior follow-up" because eighteen of the senior participants are students who participated four years earlier as freshmen in the freshman study. Of the original 32 subjects in the freshman semester study, 27 remained in engineering four years later in 1997 and eighteen agreed to participate in the experiment a second time. In addition to these eighteen follow-up senior subjects, forty-three other seniors were recruited to complete the design tasks. Thus, the data for this study consists of the responses of the original thirty-two freshman complemented by the new responses of sixty-one seniors, eighteen of whom are the follow-up subjects. In order to ensure consistency in the analysis, all of the freshman data is being re-segmented, re-coded, and re-evaluated for quality along with the data from the senior subjects.

As before, a central portion of the analysis is focused on characterizing the design process (Turns and Atman 00). Based on earlier results, we are exploring differences in exerted effort, numbers of transitions between design steps, and the extent of progression to later steps of the design process. Given the nature of the data set, we are able to test these hypothesis for the between-subject data (32 freshman and 61 seniors) and for the within-subject data (18 students). We are also able to test for pre-test effects, by comparing the data of the forty-three new subjects with the data of the eighteen follow-up subjects. Finally, because the sixty-one seniors represent a number of different majors, we plan to explore potential differences across majors.

With lessons learned from previous studies, we are also able to explore the data with more sophistication. For example, current work with data from the Midwest Floods problem is focusing on freshman-senior differences in problem definition. In particular, the use of two different codes (type of knowledge referenced and problem component addressed) is permitting an exploration of whether seniors have problem definitions that are (1) more comprehensive, (2) include more technical issues, and (3) include more global and societal issues (Bogusch et al. 00).

4 Discussion

In the introduction of this chapter, we argued that in order to improve engineering education, we need to better understand what engineering students know relative to design. The studies surveyed in this paper represent our cumulative efforts to use verbal protocol analysis as a means of gaining detailed empirical insights into the design processes of engineering students. These studies show differences in student design performance in three different cases: 1) after the first semester of a freshman year, 2) after a short term intervention such as reading a text book, and 3) after completing an undergraduate engineering degree.

The specific findings from the three completed studies are synthesized in Table 4. In the table, the checks indicate that a significant effect was discovered. A check, enclosed in parentheses, indicates the presence of a trend but no significant effect. A dash indicates that the measure was investigated but neither a significant effect nor a trend was discovered. Any empty cell indicates that a particular measure was not used in analyzing a specific problem in a specific experiment.

While these findings contribute to the research base of knowledge on engineering student design processes, a true test of the utility of this research will be the ability of this research to translate into implications for engineering design education. In the last portion of this chapter, we reflect on some instructional implications of this research.

4.1 Challenges in Teaching and Assessment in Freshman Design

The diverse behaviors of freshman designers suggest challenges and guidelines for teaching and assessing design at the freshman level (Sheppard and Jenison 97). The freshmen design timelines in Figure 1 demonstrate three different approaches to design that freshmen may have upon entering a freshman design experience. "Problems" that students may experience include getting stuck in problem scoping and spending too much time (or too much uninterrupted time) in modeling of a single alternative solution. At the same time, some freshmen have quite expert looking design processes, potentially needing help only with the decision making and communication steps of design. In the context of freshman design education, these observations suggest the basis for assessments to characterize the incoming freshman design abilities. On the other hand, they suggest an instructional challenge. What does an instructor do with such diversity of approaches to design?

4.2 Analyses and Findings as a Basis for Classroom Activities

One challenge in teaching design is to find ways to meaningfully discuss aspects of the design process without having students engage in an entire design problem. We are currently exploring ways to turn elements of our research into engaging interactive classroom activities. For example, the information gathering results of the Playground problem have led to the development of an instructional activity to discuss the topic of problem scoping with students. In the activity, we present students with the playground problem and ask students to create a list of information needs. We then have students exchange papers and code the list they receive according to the information categories used in the study. A synthesis of the results across the class and a comparison of the class results with the study results then forms the basis for a discussion of (a) information gathering behavior, (b) what information students are responsible for and why (e.g., maintenance, liability), and (c) strategies for ensuring that the information needs of the problem are addressed during design (e.g., teamwork, case studies). Students report that they enjoy the problem and find it thought-provoking.

4.3 The Unclear Path from Findings to Instruction Implications

The findings concerning progression seem to suggest an area in which more research may be needed before the educational implications of the finding are clear. In a nutshell, the progression findings are that less experienced students progress less frequently to later stages of design such as decision making.

Table 4: Synthesis of measures used across all studies.

	1st Semester Freshman Study			Design Text Study			F-S Comparison
Categories / Measures	PP	SC	MW	PP	SC	MW	PG
Product Measures							
- Solution Quality	–	(√)		–	–		√
Process Measures							
- Number of Words Spoken	√	√		√	√		
- Time Spent							–
- Number of Transitions	√	√		√	√		√
- Transition Rate							√
- Time Spent in Decision Step							√
- Progression to Product Realization Stage of Design							√
							– 1
- Number of Alternative Solutions				(√)	(√)		
- Number of Design Criteria Considered	√	√	√	√	(√)	√	
- Number of Explicit Information Requests							√
- Number of Assumptions Made							√
- Number of Information Categories Covered							√

1 Although the number of alternative solutions considered was not significant for differentiating among freshman and seniors, this measure was correlated with solution quality.

So what are the educational implications? One conclusion might be that students need to be facilitated in getting to the decision making step and that the appropriate instructional intervention would be to create some instructional approach that helps students make decisions. Another approach would be to take a step back and ask why students may be having this difficulty. What type of content knowledge or meta-knowledge is required of a designer for him or her to effectively make design decisions? Is the nature of the problem a factor – do some designers progress to decision making on some problems but not others. Maybe the issue is one of knowing why making explicit decisions among alternatives is important. In this case, verbal protocol analysis results provide an empirical anchor to get the engineering design education community talking.

5 Conclusion

In the last decade, engineering design has become a central topic in discussions of engineering education reform, leading to a strong need to understand the processes of engineering student designers. The four verbal protocol studies reported in this paper provide insight into engineering student design processes. These studies, and others like them, can form a strong research foundation for improving design learning in the engineering curriculum.

6 Acknowledgements

This work has been supported by a variety of people and funding agencies. This research was made possible in part by National Science Foundation grants RED-9358516 and DUE-9254271 (an NSF Young Investigator award to Atman) as well as matching funds from the GE Fund, the Ford Motor Company Fund, Xerox Corporation, Westinghouse Foundation, and Lockheed Martin.

Research of this type is time intensive. The authors would like to gratefully acknowledge Karen Bursic, Robin Adams, Mary Besterfield-Sacre, Laura Bogusch, Pamela Moore, Carie Mullins, Jason Saleem, Heather Umphred, and Chris Yarsky for their contributions in the design, execution, and analyses of these studies. We would like to express our appreciation to the subjects from the University of Pittsburgh who participated in the study. And finally we would like to thank 30 undergraduate student researchers at the University of Pittsburgh and the University of Washington who helped to code and analyze the data.

References

ABET. Engineering Criteria 2000: Criteria for Accrediting Programs in Engineering in the United States (Third Edition), See http://www.abet.org/downloads.htm, download EAC Criteria for 2000-01 (Includes EC2000) PDF, 32-34, 2000.

Adams R S, and Atman C J, (1999), Characterizing engineering student design processes: An illustration of iteration, American Society of Engineering Education, Charlotte, NC.

Adams R S, and Atman C J, (1999), Cognitive processes in iterative design behavior. Frontiers in Education, San Juan, Puerto Rico.

Atman C J, and Bursic K M, (1998), Verbal protocol analysis as a method to document engineering student design processes, *Journal of Engineering Education,* April, 121-131.

Atman C J, and Bursic K M, (1996), Teaching engineering design: Can reading a textbook make a difference? *Research in Engineering Design* 7(7), 240-250.

Atman C J, Bursic K M, and Lozito S L, (1995), Gathering information: What do students do? American Society of Engineering Education, Milwalkee, WI.

Atman C J, Bursic K M, and Lozito S L, (1996), An application of protocol analysis to the engineering design process, American Society of Engineering Education, Milwalkee, WI.

Atman C J, Chimka J R, Bursic K M, and Nachtmann H N, (1999), A comparison of freshman and senior engineering design processes, *Design Studies* 20(2), 131-152.

Ausebel D P, (1968), *Educational psychology: A cognitive view*, Rinehart and Winston, New York, NY.

Bogusch L L, Turns J, and Atman C J, (2000), Engineering design factors: What do students think is important? Frontiers in Education, Kansas City, MO.

Bucciarelli L L, (1994), *Designing engineers*, MIT Press, Cambridge, MA.

Bursic K M, and Atman C J, (1997), Information gathering: A critical step for quality in the design process, *Quality Management Journal,* 4, 60-75.

Chi M T H, (1997), Quantifying qualitative analyses of verbal data: A practical guide, *The Journal of the Learning Sciences,* 6(3), 271-315.

Chimka J R, and Atman C J, (1998), Graphical representations of engineering design behavior, Frontiers in Education, Tempe, AZ.

Chimka J R, Atman C J, and Bursic K M, (1997), *Describing student design behavior.* American Society of Engineering Education, Milwaukee, WI.

Christaans H H C M, and Dorst K H, (1992), Cognitive models in industrial design engineering: A protocol study, *Design Theory and Methodology*, ASME.

Cross N H, Christiaans H H C M, and Dorst K H, (1994), Design expertise amongst student designers, *Journal of Art and Design Education,* 13(1), 39-56.

Cross N H, Christiaans H H C M, and Dorst K H, Eds. (1996), *Analyzing Design Activity*, Chichester, John Wiley & Sons.

ECSEL, (1993), *Introduction to Design Engineering Active Learning Modules: A Catalogue*, Washington, DC, ECSEL National Office.

Ennis C, and Gyeszley S W, (1991), Protocol analysis of the engineering systems design process, *Research in Engineering Design,* 3, 15-22.

Ericsson K A, and Simon H A, (1993), *Protocol Analysis*, MIT Press, Cambridge, MA.

Gero J S, and McNeill T M, (1999), An approach to the analysis of deisgn protocols, *Design Studies,* 19(1), 21-61.

Guindon R, (1990), Designing the design process: Exploiting opportunistic thought. *Human-Computer Interaction,* 5, 305-344.

James C M, Goldman S R, and Vandermolen H, (1994), The role of planning in simple digital circuit design, *paper presented at the annual meeting of the American Educational Research Association.*

Mann C R, (1918), *A Study of Engineering Education.* The Carnegie Foundation for the Advancement of Teaching.

Moore P L, Atman C J, Bursic K M, Shuman L J, and Gottfried B S, (1995), Do freshman design texts adequately Define the Engineering Design Process? American Society of Engineering Education.

Mullins C A, and Atman C J, (1994), *Freshman engineer's strategies for solving open-ended problems*, American Society of Engineering Education.

Mullins C A, Atman C J, and Shuman L J (1999), Freshman engineer's performance when solving design problems, *IEEE Transactions on Education,* 42(4), 281-287.

Ng E, and Bereiter C, (1991), Three levels of goal orientation in learning, *The Journal of the learning sciences,* 1(3 and 4), 243-271.

NRC, (1995), *Engineering Education: Designing an Adaptive System*, National Academy Press, Washington, DC.

NSF, (1995), *Restructuring Engineering Education: A Focus on Change*: *Restructuring Engineering Education*, Division of Undergraduate Education, Directorate for Education and Human Resources, National Science Foundation, Washington, DC.

Oxman R, (1995), Observing the observers: Research issues in analyzing design activity, *Design Studies,* 16(2), 275-283.

Pahl G, and Beitz W, (1996), *Engineering Design: A Systematic Approach.* London, Springer-Verlag.

Radcliffe D F, and Lee T Y, (1989), Design methods used by undergraduate engineering students, *Design Studies,* 10(4), 199-207.

Rowland G, (1992), What do instructional designers actually do? *Performance Improvement Quarterly,* 5(2), 65-86.

Schoenfeld A H, (1985), *Mathematical Problem Solving*, Academic Press, New York, NY.

Sheppard S, and Jenison R, (1997), Examples of freshman design education, *International Journal of Engineering Education,* 13(4), 248-261.

Suh N, (1990), *The Principles of Design.* Oxford University Press, New York, NY.

Sutcliffe A G, and Maiden N. A. M, (1992), Analyzing the novice analyst: Cognitive models in software engineering, *International Journal of Man-Machine Studies,* 36, 719-740.

Tang J C, and Leifer L J, (1991), An observational methodology for studying group design activity, *Research in Engineering Design,* 2, 209-219.

Todd R H, Magleby S P, Sorenson C D, Swan B R, and Anthony D K, (1995), A survey of capstone engineering courses in North America, *Journal of Engineering Education,* April, 165-174.

Turns J, and Atman C J, (2000), Characterizing engineering student design process - a within subjects verbal protocol study, Presentation at the American Society of Engineering Education National Conference, St. Louis, MO.

Turns J, Atman C J, and Sidiadinoto I, (1999), Students use of functional, behavioral and structural terms to describe artifacts during design, presentation given at the American Society of Engineering Education National Conference, Charlotte, NC.

Ullman D, Dietterich T G, and Stauffer L A, (1988), A model of the mechanical design process based on empirical data, *AI EDAM,* 2(1), 33-52.

Umphred H N, Atman C J, Turns J, and Shuman L, (in progress), Relating design process to product quality, *working paper.*

Chapter 4

Novice conceptions of design: Implications for the design of learning environments

Wendy C. Newstetter
W. Michael McCracken
College of Computing
Georgia Institute of Technology
Atlanta, GA

Abstract

Our experience in designing and teaching a cross-disciplinary freshman design class has led us to believe that students entering design fields are saddled with naïve or (mis)conceptions about design and design activity. It is our belief that for students to become effective designers, they must be helped to recognize and overcome these misconceptions through appropriate educational interventions. To better understand the nature and substance of these misconceptions, we have used several methods of studying the students in the class, and we have conducted a descriptive survey study of 290 freshman. Our findings begin to suggest a consistent profile of misconceptions across disciplines.

1 Introduction

Learning of any kind is filtered and interpreted through the learner's lived experiences. Such experiences, and the understanding accrued, constitute the learner's prior knowledge. Our hunch is that students of design have well-developed prior conceptions and theories about the nature of design that conflict with understandings held by expert designers. We imagine that such understandings arise from childhood experiences building with blocks, from constructing forts out of tablecloths and card tables, and from assembling dioramas for class projects. Student theories of design also come from media images of designed products, from fashion to cosmetics to cars. Design here is most often depicted as the creation of consumer products. In composition and writing classes, it is not uncommon for students to practice design as brainstorming. In this context, it is generating novel ideas. Conceptions of design are also crafted from years of school problem solving where there is one sanctioned route and one answer to the problem. Students routinely solve design problems as though they were algorithms to be parsed.

Whatever the source of these prior conceptions, they are often in conflict with the knowledge and practices of expert designers. Of critical importance to design researchers and educators, therefore, is discovering and understanding the nature of learner conceptions of design as well as their origins. Successful teaching and learning strategies are contingent upon knowing what the learner of any subject or discipline brings to the classroom environment, for prior knowledge can either facilitate or grossly retard new learning. Towards that end, we have collected various types of data from undergraduate learners of design at a large technological institute. Our preliminary findings suggest features of prior conceptions that have important implications for the design of learning environments. This paper reports on those findings and suggests further avenues of research needed to better understand the filters and interpretations that often impede design learning.

2 Learning as conceptual change

Cognitive science research has shown that a learner's prior knowledge is critical to the development of an integrated and generative knowledge (Chi and Ceci 87; Chi, Glaser and Farr 88; Glaser 84). As Glaser and deCorte explain, "A well-organized and coherent knowledge base initiates inference, conceptualization, and the acquisition of principled understanding" (Glaser and De Corte 92). It seems a certainty that prior knowledge is an essential variable in design learning. Unfortunately, however, prior knowledge is often incorrect. Variously referred to as *preconceptions* (Clement 82) *naïve theories* (McClosky 83)' *alternative frameworks* (Driver 83) and *alternative theories* (Resnick 83), these incorrect notions based on experiences coalesce into knowledge structures that operate in theory-like ways. As such, they have explanatory power, inform action and often resist change even when there is repeated instructional evidence that the stored information is faulty.

As stated above, a learner's experiences often coalesce into a theory-like framework. A theory, as cognitive scientists conceive of it, operates as a representational system. Johnson-Laird (Johnson-Laird 83) describes a theory as manifesting itself in three forms: 1) "propositional" representation—syntactic strings of symbolic elements (e.g. Constraint setting is a part of design) 2) "mental models"—structural analogs of real-world or imagined situations (e.g. A designer is setting constraints) 3) images—a mental model from a particular perspective (e.g. The architect is constraining the problem). Put another way, these representations represent the "what" of the theory, "the how" of the theory and the "under what circumstances" of the theory. Theories are valuable because they shorten the interpretation and meaning-making process considerably by doing the inferential work. Humans readily develop theories to explain phenomena and cope with constant stimuli that might otherwise force them to attend more closely.

Following Piaget, who demonstrated so clearly how children and adults differ, educational researchers in the late 1970s started to probe these learner theories and supporting knowledge structures, particularly in the areas of mathematics and science learning, the goal being to explain the documented difficulties students have in fully grasping scientific explanations for things like bodies in motion and electrical circuitry (Chi 92). Since that time numerous studies have demonstrated how such misconceptions can impede effective learning. Most agree that for learning to occur, at least replacement, often restructuring, and possibly even demolition of the knowledge base, has to occur. What is not agreed upon is the meaning of restructuring or the means to accomplishing it. But before tackling the nature of or means to restructuring a base of knowledge, it is important to interrogate the structure of knowledge itself.

Chi (98) has argued that there are three types of misconceptions: *incorrect, inconsistent* or *incompatible.* Incorrect misconceptions are relatively easy to change because they are fragments or random collections of false ideas. In the Johnson-Laird framework, they would be incorrect propositions. In contrast, inconsistent prior beliefs resist change because they are part of a larger mental model that has structure. Learning in this scenario involves not only replacement of false beliefs with accurate ones, but also the dismantling of mental structures that can be robust although inaccurate. Students who have developed responses to design problems that mimic responses to well-structured algorithmic problems could be seen as operating from an inconsistent set of beliefs. When prior knowledge is incompatible with expert understanding, the most radical form of conceptual change is required because such knowledge is highly resistant to change. In such cases, novice understanding can best be comprehended as a fully structured paradigm with explanatory powers and reflexively constructed false explanations for external phenomena. When confronted with anomalies to the theory, they can be rationalized away within the incompatible framework because the naïve understanding operates much like a fully developed theory. Such forms of knowledge are extremely difficult to change. In fact, Chi has suggested that they are never fully eradicated but exist alongside the more expert version, somewhat dormant but at times available for faulty explanation. This framework for understanding the nature of prior knowledge is depicted in Figure 1 below.

The implications for pedagogy are clear. The more *incommensurable* the set of misconceptions is to that of an expert, the more aggressive the learning intervention needs to be. This has been demonstrated repeatedly in science learning where naïve theories are remarkably robust and resistant to change regardless of the pedagogy.

Prior Knowledge

Incorrect	**Inconsistent**	**Incompatible**
Piecemeal; false beliefs	wrong mental model, robust & structured	ontological misconceptions robust/ reflexive interpretation
Removed through refutation/contra diction	Removed through confrontation between MM/ continuous & gradual self-repair	Removed through paradigm shift

Figure 1: Chi, M. Talk at Georgia Tech 1998

3 Prior knowledge and misconceptions in design learning

In our work at Georgia Tech, we make the assumption that every student brings a naïve theory of design to the classroom that might be partially generalizable across students and partially idiosyncratic. Our hunch is that these conceptions are inconsistent, which that they are anchored in a mental model that has a certain robustness. It also means that our pedagogic practices must bring about a confrontation between the learner model and the expert model. Having students follow prescriptive models of design, we believe, does not constitute confrontation of the sort that can begin the dismantling of the mental model. Students are masters of following teacher tasks with out learning them as ways of doing (Newstetter 98), just like science learners can parrot back the laws governing bodies in motion on a test but cannot explain why an object moves as it does in real-time. We need to develop means to probe learner misconceptions to better understand what we are up against. Will learning entail the addition of new propositional elements, the restructuring of existing elements into a new structure or the eradication of existing elements and radical replacement? Each of these possible learning scenarios suggests very different types of interventions.

At present we do not have systematic descriptions or accounts of learners' theories of design. However, verbal protocol studies of novice and expert designers clearly indicate that the activities or design processes of these groups differ. In a study of industrial design engineers, Christiaans and Dorst (Christiaans and Dorst 92) found that novices tend to scope out the problem less and seek less information than experienced designers. At the same time, however, they decompose the problem into more solvable parts without attention to the whole design. Consistent with these findings, Rowland (Rowland 92) found that novices interpreted design problems as well-defined and as a result, did little elaboration to explore and close the design space. In another study of novices, Sutcliffe and

Maiden (Sutcliffe and Maiden 92) found that while experts reason with conceptual models and rigorously test out hypotheses, novices fail to generate hypotheses or use models for reasoning. Atman et al. (Atman, Chimka, Bursic and Nachtmann 99) found in their study of university freshman and senior designers that seniors asked for significantly more information in the problem-scoping phase, transitioned more frequently and faster between design steps and spent much longer in decision-making at the product realization stage than freshman.

Our research over the last three years has sought to uncover and begin to describe these naïve conceptions of design through 1) observation in our design learning laboratory (DLL) 2) collection of learner concept maps in the DLL and 3) a survey administered to 520 students in an introductory computer science course. In the next three sections we report on preliminary findings from those efforts.

4 Findings from classroom observation

The Design Learning Laboratory at Georgia Tech seeks to better understand the nature of design learning and the most effective means to apprentice that learning. Our activities include collecting data on novice conceptions of design using observational, qualitative and quantitative studies, developing classroom interventions based on those conceptions and testing those interventions in the classroom. This means we conduct local, in-class experiments or what Brown (Brown 92) calls "design experiments" situated in real classrooms, and undertake empirical studies to clarify the nature of learning that occurred and its relationship to the intervention.

Our observations of student behavior over the last three years as evidenced in their work, their classroom accounts and their reflective journals have yielded the following characterizations of typical student behavior. These characterizations have been abstracted away from the products and processes demonstrated by the students. Each describes a feature of novice design activity as we have seen t play out over eight school terms.

1. *Ideation without substance* - Students believe design is coming up with good ideas. Obviously design has ideation as a constituent element but designers also concern themselves with the realizability of ideas and evaluate ideas based on informed decision-making and analysis.
2. *Design arrogance* - Students do not place their designs in the context of the environment in which the design will reside. They "arrogantly" ignore the constraints of the user (whether that is a machine or a person). They often design for themselves.
3. *Design shutdown* - Students tend to focus on single point solutions to problems once beyond the ideation stage. In other words, once they have an idea, they stop considering alternatives and focus all their energy on that one solution regardless of its feasibility.

4. *Design jumps* -Students have a tendency to operate at only two levels of abstraction. The highest level of general ideas (function), and the lowest level of the components of the product (structural). They do not move between these spaces in any formal manner, nor do they consider the ramifications of the giant leaps they are taking between those two levels of abstraction.
5. *Design routinization*- Students act as though designing is a serial/ linear process. The way they deal with design problems resembles the linear parsing of the algebra problem. Iteration, revisiting past decisions and evaluating alternatives is not in their process model.

These general observations motivated us to try to gather data on the prepositional and relational conceptions learners had when entering the classes in the lab. Developing instruments to gather this kind of data is difficult. In the next sections we present some particularly compelling data gathered using the technique of concept maps.

5 Findings from concept maps

In the Design Learning Lab at the start of every term, we collect learner concept maps of design as a source of data for developing our understanding of learner misconceptions. The directions ask them to first lit all the words they associate with the process of design. We next ask them to draw a representation of the design process using the words they have written down. The meaning of representation is left vague to allow them to construct what seems appropriate to them. In most cases, students draw linear flow charts. But some have been different in their creativity. We have found that these renderings of design are prophetic in anticipating the kinds of behaviors we have observed in students discussed above. Two particularly insightful ones are featured below.

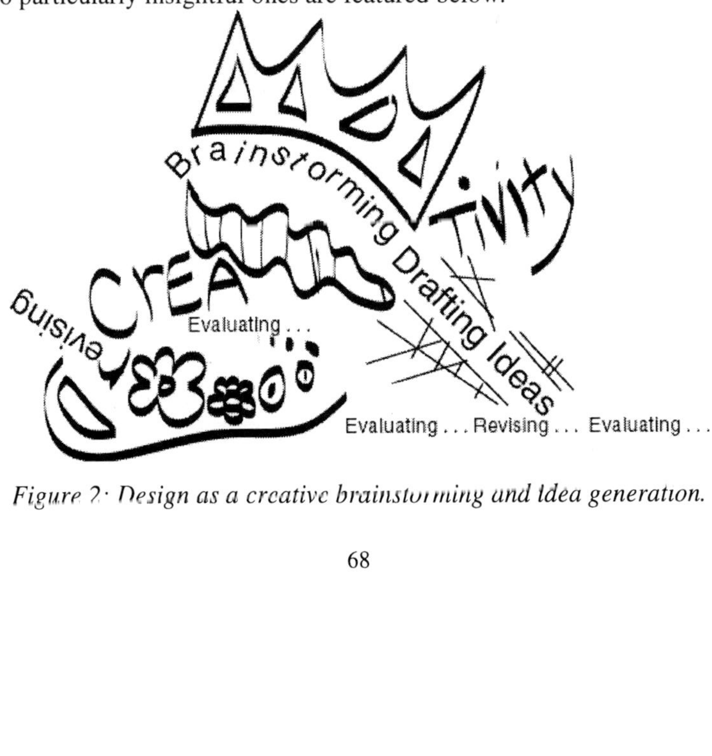

Figure 2: Design as a creative brainstorming and idea generation.

Figure 2 constructs design as creative brainstorming and idea generation. Evaluating and revising are peripheral to the central, larger than life activity of creating ideas. Using Johnson-Laird's notion of theory, we could say that this student's propositional representation of design is quite impoverished. Design consists of only five elements: creativity, brainstorming, drafting ideas, evaluating and revising. The mental model foregrounds the role of creativity while diminishing the importance of evaluation and revision. This naïve theory mirrors the popular image of design being a stroke of genius, the gift of a talented few. It is an image from the arts, and not without credibility where creative genius is the hallmark of design success. One problem (of many) with this theory is that creativity holds up the whole house of cards; without it, design fails.

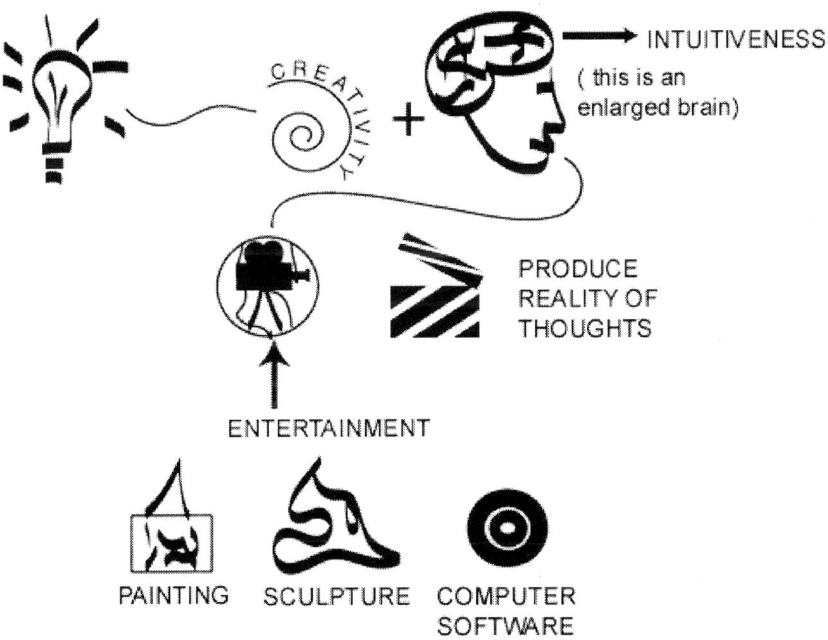

Figure 3: Centrality of the creative process

In Figure 3 we again see the centrality of the creative process. It entails three steps that magically produces "reality of thoughts" which materialize unto various kinds of products. Propositionally we see five elements: imagination, creativity, intuitiveness, thoughts and products (of various kinds). The mental model suggested by the depiction hints at levels or phases in the process, the first in the head, the next somewhere in reality and the last as marketable products. The popular image of the illuminating light bulb for imagination renders design as that blaze of creative light that strikes some and not others. The symbols at the bottom begin to suggest Johnson-Laird's image representations where design occurs in the arts and in computer software. Intriguing as these images of design are, what implications do they have for the goal of this paper?

6 Findings from a survey study

After analyzing the data from the studies above, we determined that we needed to verify emerging findings and themes with a much larger population. We therefore decided to undertake a large-scale survey study based on findings from the earlier data. We identified the introductory computer science course as a logical site for this study as every student in the institution is required to take this course and findings could be generalizable to a greater population rather than a group of self-selected students in the DLL. The goal of the survey was to determine if misconceptions of designing existed in freshman engineering and computer science students. The survey was designed to illuminate the misconceptions.

Designing an instrument that attempts to uncover misconceptions is difficult. We employed three techniques in the survey to uncover the misconceptions, if they existed. We employed free form questions to allow the subject to describe design and designing in their own terms. We used terminological ranking techniques to: 1) Probe the student's understanding of the terms of designing and 2) Uncover their perceptions of design activities. The third technique asked the students to compare two design problems and their solutions. The intent of that comparison was to determine if they could differentiate good design processes from bad design processes. The survey was administered on-line to 520 students who were taking CS1501, Introduction to Computing. The course is required of all CS majors as well as many of the engineering programs at Georgia Tech. We had independent expert review of the survey, and ran a small pilot of the survey prior to administering it to the full population. The survey population represented most engineering and computing students at our university, and included a cross-section of freshman through seniors. We concentrated our initial analysis to the freshman to isolate variables of acquisition of design skill through their programs. Appendix A contains the complete survey.

6.1 Analysis of Survey Data

520 students responded to the survey. We analyzed the responses from the freshman students (290 respondents). The analysis included extraction of common themes from the free form responses and quantitative analysis of the choice responses. This chapter concentrates on five of the questions. The first, examples of designing, was constructed to allow them to describe designing in their own terms. We coded their responses into categories and generated category counts to see if there was any clear trend or dominant

category. The next two questions we analyzed were a pair of questions asking them to rank from a list the five most important and five least important terms describing designing. The responses were tested for significance and ranked as a function of response. The last two questions we analyzed asked them to independently rank the quality of the design process of two design problems. We analyzed their responses to determine if the participants were able to differentiate the good design process from the bad design process.

6.2 Examples of Designing

Question: *List five examples of what comes into your head when you think of designing.*

Our analysis of the data showed student responses falling into three categories. These included, product response, e.g., programs, fashion, web pages, or planes; domain response, e.g., mechanical design, interior design; or design activity, e.g., invention, or planning. Our coding scheme allowed for duplicate scoring if the student included product, process, and/or activity descriptions. The coding was not rater reliability tested. The following table summarizes the results of coding the responses.

Table 1: Coding of responses.

Response Type	Response Score
Product	140
Domain	163
Activity	77

The results of this portion of the data offer several insights. As we expected, many of the students thought of design as products. In a culture that promotes commercial endeavors, we would expect answers to be informed by media driven notions of design. In contrast, those students who thought of designing as a process, primarily thought of it as a domain description, such as interior design, fashion design, architecture, or graphics. Interestingly, although they are at an engineering institute, few of the respondents included engineering design in their lists, which may indicate they don't think of this as a design discipline. Finally, those who described design as activities generally only included activities such as brainstorming, ideation, or creativity. Few responded with normally accepted descriptions of design activities such as iteration, evaluation, planning, etc. The result is of concern as it relates to misconceptions. We find it significant that very few thought of engineering as design, and very few listed what expert designers would consider critical design activities in a free choice question. Finally, many of the respondents listed programming, and not program design or software design. We expected software type of responses since this is a computer science class, and if our beliefs of misconceptions are true, then the programming response is appropriate. In other words, the students believe design and synthesis are synonymous.

6.2.1 Relevance and Lack of Relevance of Terms in Designing

Two more questions in our survey attempted to probe student misconceptions by asking them to rank terms that describe design activities. The first question asked the students to rank from a list of sixteen terms the five they felt most accurately described designing. Similarly, we asked the students to rank from the same list the five terms they felt least accurately described designing. Each set of responses was tested for significance ($p<.05$) using a chi-squared goodness of fit test. The tables below summarize the answers from those two questions.

Table 2: Top five most important terms describing designing

Ranking	Term Describing Design
1.	Understanding Problem
2.	Using Creativity
3.	Visualizing
4.	Brainstorming
5.	Making Decisions

Table 3: Top five least important terms describing designing

Ranking	Term Describing Design
1.	Making Trade-offs
2.	Decomposing
3.	Synthesizing
4.	Generating Alternatives
5.	Sketching

We find it interesting and perhaps predictable that terms such as brainstorming, creativity, and visualizing are not activities discussed in most design process descriptions, see for example (Goel and Pirolli 92). We suspect that these notions of design are carried over from high school activities related to writing instruction and other creative activities. Although these are important in design, they are generally not considered the critical activities of effective design. What was most enlightening was the students responses to the least relevant terms, as they are generally included in most design process descriptions as critical.

These findings, in particular the least relevant responses, begin to explain our earlier observations of how misconceptions manifest themselves in novice design activity. As an example, design shutdown could be explained as not understanding the value of generating alternatives, or possibly design arrogance could be explained as originating as failing to understand the decomposing of problems.

6.2.2 Design Problem Analysis

We asked the students to review two problems that characterized simple design processes, and score the problems on a scale as to whether the activities were "good designing". The first problem characterized designing in the form of our misconceptions (routinization, shutdown, ideation). The second problem characterized designing in the more accepted form (generation and evaluation of alternatives, iteration, etc.). The results of the students scoring of the problems was tested for significance to ensure the two problems were treated independently by the respondents. A paired sample t-test was run on the data with the results of that test in the following table which shows that the two answers were significantly different.

6.3 t-Test:Paired Two Sample for Means

	Problem 1	Problem 2
Mean	6.25172414	5.23103448
Variance	3.72534304	3.67308197
Observations	290	290
P(T<=t) two-tail	5.4545E-10	

Though statistically the data reflects little difference between the two responses, practically the results are quite interesting. The students were unable to recognize the differences between the two problems and were interestingly neutral in their scores for both. We can conjecture that the students observed the problem from a solution view. The problems and their solutions were purposely designed to be unexciting in their presentation to minimize the bias of the students picking the "coolest" design. Just as importantly, it is likely that they didn't perceive any differences between the two design scenarios, or they were unable to judge one better than the other. The fact that the judgements about both centered on the middle of the scale indicates indecision and uncertainty, whereas, data congregating on either the high or low end would have indicated a commitment to one design scenario or the other. This indicates to us the students' inability as novices to evaluate effective design processes from ineffective ones.

These results substantiate some of the findings from our classroom observation and the student's concept maps. The concept maps reflect student's views of designing as creativity, brainstorming, invention and flashes of brilliance, and that design was not an activity, but a product or product domain. Our data on student's free form responses as well as their rankings of terms correlates with the observations. The students see design as possibly unrelated to their field. In other words, engineers don't design, they engineer (whatever that is) and computer scientists program. Their responses imply that design is the world of invention or the arts. The data we analyzed on their ranking of terms followed the same path as our observations of their views of design as an activity. They ranked as least important terms that describe design activities and ranked as most important the creative elements of design. Our observations of student behavior in our classes are commensurate with these findings. Design shutdown, ideation without substance, and design jumps can all be attributed to their beliefs that designing is primarily coming up with good ideas. In the classroom we have also observed their attempts at serialization of the design process. Terms like iteration and evaluation (as we asked them to rank in the survey data) are foreign to them in the context of design. We now know at least some of the misconceptions of novice designers. We have observed and empirically tested our observations. We don't claim our data is valid or infallible but we do believe misconceptions are impeding learning to design. Our concerns now are how deep-seated these misconceptions are and what interventions are necessary to unseat them.

7 Conclusions

To be able to better integrate design throughout a design curriculum, a deeper understanding of students' naïve conceptions of design is required. Our goal is to better understand the actual substance of students' misconceptions of design. The preliminary results reported in this paper confirm our initial beliefs that we can begin to develop a catalog of students' misconceptions of designing and prescribe interventions to overcome those misconceptions. With this work we have just begun to understand the misconceptions students may harbor. We have a preliminary study that supports our observational data, yet many questions remain unanswered. Our experience with novice designers leads us to believe that understanding naïve conceptions of design is critical in transitioning students to expert design activity. Thus, our misconceptions research must still answer the following questions:

1. Are (mis)conceptions consistent across ages and populations?
2. Are (mis)conception theory-like or piecemeal?
3. How intractable the mis(conceptions)?
4. What are the sources of (mis)conceptions?

But we believe there is more to learning designing than just overcoming misconceptions. The answers to the questions on misconceptions will generate a critical part of an overall framework for the science of design learning. The framework of the science of design learning's goals are: 1) understand how experts practice design, 2) understand how people learn to design and what makes learning to design difficult, and 3) what are the classroom practices that need to be developed to effectively educate designers? With a framework of design learning we can begin to forge strong links between research into the science of design, the development of design pedagogy and a robust theory of design learning.

References

Atman C J, Chimka J, Bursic K, and Nachtmann H, (1999), A comparison of freshman and senior engineeirng design processes, *Design Studies,* 20(2), 131-152.

Brown A, (1992), Design experiments: Theoretical and methodological challenges of creating complex interventions in classroom settings, *Journal of Learning Sciences,* 2(2), 141-178.

Chi M T, and Ceci S J, (1987), Content knowledge: Its role, representation, and restructuring in memory development, H. W. Reese Ed. *Advances in child development and behavior*, Orlando, FL, Academic Press, 20, 91-142.

Chi M T, Glaser R, and Farr M, (1988), *The nature of expertise*, Lawrence Erlbaum and Associates, Hillsdale, NJ.

Chi M T H, (1992), Conceptual change within and across categories: Implications for learning and discovery in science, R. Geere Ed, *Cognitivie models of science, Minnesota studies in the philosophy of science*, 15, University of Minnesota Press, Minneapolis, MN.

Christiaans H H C, and Dorst K H, (1992), Cognitive models in industrial design engineering, *Design theory and methodology,* 42, 131-140.

Clement J, (1982), Students' preconceptions in introductory mechanics, *American Journal of Physics* (50), 66-71.

Driver R, (1983), *The pupil as scientist?* Milton Keynes: Open University Press.

Glaser R, (1984). Education and thinking: The role of knowledge, *American Psychologist, 46*, 1-46.

Glaser R, and De Corte E, (1992), Preface, F. J. R. C. Dochy Ed. *Assessment of prior knowledge as a determinant for future learning,* Utrecht/London, Lemma, B.V./Jessica Kingsley Publishers 1-2.

Goel V, and Pirolli P, (1992), The structure of design problem spaces, *Cognitive Science, 16*, 395-429.

Johnson-Laird P N, (1983), *Mental Models*, Harvard University Press, Cambridge.

McClosky M, (1983), Naive theories of motion. D. Gentner and A. L. Stevens Eds. *Mental models*, Lawrence Erlbaum and Associates, Hillsdale, NJ.

Newstetter W, (1998), Of green monkeys and failed affordances: A case study of a mechanical engineering design course, *Research in Engineering Design,* 10, 118-128.

Resnick L B, (1983), Mathematics and science learning: A new conception, *Science, 220*(477-478).

Rowland G, (1992), What do instructional designers really do? *Performance Improvement Quarterly, 5*(2), 65-86.

Sutcliffe A G, & Maiden, N. A. M. (1992), Analyzing the novice analyst: Cognitive models in software engineering. *International Journal of Man-machine Studies,* 36, 719-740.

Appendix A

Survey Instrument

1. What is your major?

2. What design experience(s) have you had?

3. When you think of the words design or designing, what images pop into your head? List the first five examples of designing that you think of.

4. What kinds of activity do you think go into design? List 10 words you associate with design activity.

5. Of the sixteen design activities below, put a check mark next to the five most important.

1.	Making decisions	9.	Synthesizing
2.	Understanding the problem	10.	Visualizing
3.	Using creativity	11.	Sketching
4.	Abstracting	12.	Imagining
5.	Goal Setting	13.	Making trade-offs1
6.	Building	14.	Brainstorming
7.	Evaluating	15.	Generating Alternatives
8.	Decomposing	16.	Modeling

6. Of the sixteen design activities below, put a check mark next to the five least important.

1. Making decisions
2. Understanding the problem
3. Using creativity
4. Abstracting
5. Goal Setting
6. Building
7. Evaluating
8. Decomposing
9. Synthesizing
10. Visualizing
11. Sketching
12. Imagining
13. Making trade-offs1
14. Brainstorming
15. Generating Alternatives
16. Modeling

7. Given the following problem solution, evaluate it on a sliding scale of 1-10 as whether it is an example of a good design process, 10 being the best.

___ Jim is tasked with designing a new disk carrier for students. Jim starts solving the problem by understanding the design of disks, and then observing students current methods of transporting disks. He also goes to stores, and cruises the web looking for examples of disk carriers. At this point, he starts generating ideas for disk carriers, and comes up with what he thinks is a good carrier for students. The carrier is like the contemporary insulated nylon lunch boxes that are popular. His carrier can accommodate floppies and zip disks, is relatively cheap, and will fit in back packs.

8. Given the following problem solution, evaluate it on a sliding scale of 1-10 as whether it is an example of a good design process, 10 being the best.

___Mary is tasked with designing the layout for her dorm room. She starts by sketching the room, and cuts out scale models of the furniture of the room. She then tries multiple versions of the room layout using her model until she is satisfied that she has arrived at the optimal solution of the layout. At that point, she enlists the aid of some males (strong back, weak mind), to reorganize the actual furniture of the room.

Chapter 5

Design Cognition: Results From Protocol And Other Empirical Studies Of Design Activity

Nigel Cross
Department of Design and Innovation
The Open University
Milton Keynes, UK

Abstract

The paper reviews protocol and other empirical studies of design activity, and summarises results relevant to understanding the nature of design cognition from an interdisciplinary, domain-independent overview. Results are presented in three major aspects of design cognition - the formulation of problems, the generation of solutions, and the utilisation of design process strategies. Parallels and comparisons between results are drawn, and a number of issues identified. Many similarities of design cognition across domains of professional practice are found. It seems that the 'intuitive' behavior of experienced designers is often highly appropriate to design tasks, although appearing to be 'unprincipled' in theory.

1 Introduction

For thirty years there has been a slow but steady growth in empirical research studies of design cognition - the pioneering work being the study of architects by Eastman (69, 70). A substantial and varied range of research methods has been adopted and adapted for the investigation of design activity, including the following:

- *Case studies* have usually focused on one particular design project at a time, with observers recording the progress and development of the project either contemporaneously or post-hoc. Both participant and non-participant observation methods have been included, and both real and artificially-constructed design projects have been studied.

- The more formal method of *protocol studies* has usually been applied to artificial projects, because of the stringent requirements of recording the protocols - the 'thinking-aloud' and associated actions of subjects asked to perform a set design task. Both inexperienced (usually student) designers and experienced designers have been studied in this way.

- There have also been a few examples of *performance tests* conducted under controlled, laboratory conditions, in which subjects are required to perform a specialised task, and data on their performance is recorded and analysed. The models for these kinds of tests are the controlled laboratory studies of psychology research.

YEAR	ARCHITECTURE	INDUSTRIAL DESIGN	MECHANICAL ENGINEERING	ELECTRONIC ENGINEERING	SOFTWARE DESIGN	VARIOUS/ OTHER
1970	Eastman					
1973	Foz					
1978	Akin					
1979	Lawson					Thomas
1981					Jeffries	
1983	Schön					
1985					Adelson	
1987		Ballay	Stauffer			
1988	Eckersley		Ullman			
1989			Radcliffe			
1990	Chan		Tang Visser		Guindon	
1991	Goldschmidt		Jansson Purcell	Colgan		
1992		Christiaans	Ehrlenspiel		Davies Olson	Goel
1993			Fricke			
1994			Lloyd	Ball		
1995		Dorst				
1997	Suwa					Göker
1998		Valkenburg		McNeill		Smith
1999			Atman			

Figure 1: Summary Chart - Protocol and Other Formal Studies of Design Activity

Of all the empirical research methods for the analysis of design activity, protocol analysis is the one that has received the most use and attention in recent years. It has become regarded as the most likely method (perhaps the only method) to bring out into the open the somewhat mysterious cognitive abilities of designers. It was given a significant boost by the Delft Design Protocols Workshop of 1994 (Cross et al. 96). It does, though, have some severe limitations, to be noted later.

The chart above summarises the history of the development of this field of the formal study of design activity. This is not an exhaustive chart of all the studies that have been conducted, but is designed merely to indicate the range and timescales of developments, across different design domains.

Although the amount of research in design activity has grown substantially since the mid-1980s, the total amount still is not particularly great, and the results of that research are varied, often based on single or small numbers of subjects, and usually untested by repeat studies. Nevertheless, in this paper I propose to try to pick out some consistent patterns that may be discerned in those results, and to identify issues that are pertinent to the utilisation of the results (for example, in design education) and to further research. I will take a cross-disciplinary, or domain-independent view of the field, and try to integrate results from studies across the various domains of professional design practice.

In analysing design cognition, it has been normal until relatively recently to use language and concepts from cognitive science studies of problem solving behavior. However, it has become clear that designing is not normal 'problem solving' (Cross 99). We therefore need to establish appropriate concepts for the analysis and discussion of design cognition. For example, designing involves 'finding' appropriate problems, as well as 'solving' them, and includes substantial activity in problem structuring and formulating, rather than merely accepting the 'problem as given'. The first main area in which I will present my interpretations of findings, patterns and issues in design cognition is therefore that of how designers *formulate problems*. The second main area will be how designers *generate solutions*, since that is the over-riding aim and purpose of design activity: to generate a satisfactory design proposal. And the third main area will be the *process strategies* that designers employ, because there has been a lot of interest in design methodology - the understanding and structuring of design procedures - especially in the context of design education.

2 Problem Formulation

It is widely accepted that design 'problems' can only be regarded as a version of ill-defined problems. In a design project it is often not at all clear what 'the problem' is; it may have been only loosely defined by the client, many constraints and criteria may be un-defined, and everyone involved in the project may know that goals may be re-defined during the project. In design, 'problems' are often defined only in relation to ideas for their 'solution', and designers do not typically proceed by first attempting to define their problems rigorously.

One of the concerns in some other areas of design research has been to formulate design problems in *well*-defined ways. This is intended to overcome some of the inherent difficulties of attempting to solve ill-defined problems. However, designers' cognitive strategies are presumably based upon their normal need to resolve ill-defined problems. Thomas and Carroll (79) carried out several observational and protocol studies of a variety of creative problem solving tasks, including design tasks. One of their findings was that designers' behavior was characterised by their treating the given problems *as though* they were ill-defined problems, for example by changing the goals and constraints, even when they could have been treated as well-defined problems. Thomas and Carroll concluded that: 'Design is a type of problem solving in which the problem solver views the problem or acts as though there is some ill-definedness in the goals, initial conditions or allowable transformations.' The implication is that designers will be designers, even when they could be problem-solvers.

2.1 Goal analysis

This 'ill-behaved' aspect of design behavior has been noted even from the very earliest formal studies. Eastman (70), in the earliest recorded design protocol study (of architectural design), found that: 'One approach to the problem was consistently expressed in all protocols. Instead of generating abstract relationships and attributes, then deriving the appropriate object to be considered, the subjects always generated a design element and then determined its qualities.' That is to say, the designer-subjects jumped to ideas for solutions (or partial solutions) before they had fully formulated the problem. This is a reflection of the fact that designers are solution-led, not problem-led; for designers, it is the evaluation of the solution that is important, not the analysis of the problem.

It is not just that problem-analysis is weak in design; even when problem goals and constraints are known or defined, they are not sacrosanct, and designers exercise the freedom to change goals and constraints, as understanding of the problem develops and definition of the solution proceeds. This was a feature of designer behavior noted by Akin (78) from his protocol studies of architects: 'One of the unique aspects of design behavior is the constant generation of new task goals and redefinition of task constraints.' As Ullman et al. (88) pointed out, from studies in mechanical engineering, only some constraints are 'given' in a design problem; other constraints are 'introduced' by the designer from domain knowledge, and others are 'derived' by the designer during the exploration of particular solution concepts.

The formulation of appropriate and relevant problem structures from the ill-defined problem of a design brief is not easy - it requires sophisticated skills in gathering and structuring information, and judging the moment to move on to solution generation. Christiaans and Dorst (92), from protocol studies of junior and senior industrial design students, found that some students became stuck on information gathering, rather than progressing to solution generation. Interestingly, they found that this was not such a significant difficulty for junior students, who did not gather a lot of information, and tended to 'solve a simple problem', being unaware of a lot of potential criteria and difficulties. But they found that senior students could be divided into two types. The more successful group, in terms of the creativity quality of their solutions, 'asks less information, processes it instantly, and gives the impression of consciously building up an image of the problem. They look for and make priorities early on in the process.' The other group gathered lots of information, but for them 'gathering data was sometimes just a substitute activity for actually doing any design work' (Cross et al. 94).

A similar finding was reported by Atman et al. (99), who found from their protocol analysis studies of engineering students that, for novices (freshmen with no design experience), '. . those subjects who spent a large proportion of their time defining the problem did not produce quality designs.' However, with senior students, Atman et al. did find that attention to 'problem scoping' (i.e. 'adequately setting up the problem before analysis begins', including gathering a larger amount and wider range of problem-related information) did result in better designs. As with the industrial design students, some of the freshmen engineering students, it seemed, simply became stuck in problem-definition and did not progress satisfactorily into further stages of the design process.

2.2 Solution focusing

Many studies suggest that designers move rapidly to early solution conjectures, and use these conjectures as a way of exploring and defining problem-and-solution together. This is not a strategy employed by all problem-solvers, many of whom attempt to define or understand the problem fully before making solution attempts. This difference was observed by Lawson (79), in his experiments on problem solving behavior in which he compared scientists with architects: 'The scientists were [attempting to] discover the structure of the problem; the architects were proceeding by generating a sequence of high-scoring solutions until one proved acceptable... [The scientists] operated what might be called a problem-focusing strategy... architects by contrast adopted a solution-focusing strategy.' Lawson repeated the experiment with first-year students of science and architecture, and did not find the same difference in problem solving strategy. It appears to be a difference that grows with education and experience in designing.

Lloyd and Scott (94), from protocol studies of experienced engineering designers, found that a solution-focused approach appeared to be related to the degree and type of previous experience of the designers. They found that more experienced designers used more 'generative' reasoning, in contrast to the deductive reasoning employed more by less-experienced designers. In particular, designers with specific experience of the problem type tended to approach the design task through solution conjectures, rather than through problem analysis. They concluded that 'It is the variable of specific experienceof the problem type that enables designers to adopt a conjectural approach to designing, that of framing or perceiving design problems in terms of relevant solutions.'

2.3 Co-evolution of problem-solution

Designers tend to use solution conjectures as the means of developing their understanding of the problem. Since 'the problem' cannot be fully understood in isolation from consideration of 'the solution', it is natural that solution conjectures should be used as a means of helping to explore and understand the problem formulation. As Kolodner and Wills (96) observed, from a study of senior student engineering designers: 'Proposed solutions often directly remind designers of issues to consider. The problem and solution co-evolve.'

This interpretation of design as a co-evolution of solution *and* problem spaces has also been proposed by others, and has been found by Cross and Dorst (98) in protocol studies of experienced industrial designers. They reported that: 'The designers start by exploring the [problem space], and find, discover, or recognise a partial structure. That partial structure is then used to provide them also with a partial structuring of the [solution space]. They consider the implications of the partial structure within the [solution space], use it to generate some initial ideas for the form of a design concept, and so extend and develop the partial structuring. . . They transfer the developed partial structure back into the [problem space], and again consider implications and extend the structuring of the [problem space]. Their goal . . . is to create a matching problem-solution pair.'

2.4 Problem framing

Designers are not limited to 'given' problems, but find and formulate problems within the broad context of the design brief. This is the characteristic of reflective practice identified by Schön (83) as problem setting: 'Problem setting is the process in which, interactively, we *name* the things to which we will attend and *frame* the context in which we will attend to them.' This seems to characterise well what has been observed of the problem formulation aspects of design behavior. Designers select features of the problem space to which they choose to attend (naming) and identify areas of the solution space in which they choose to explore (framing). Schön (88) suggested that: 'In order to formulate a design problem to be solved, the designer must *frame* a problematic design situation: set its boundaries, select particular things and relations for attention, and impose on the situation a coherence that guides subsequent moves.'

This kind of problem framing has been noted often in studies of architects. Lloyd and Scott (95), from their studies of (mostly senior-student) architects, reported that 'In each protocol there comes a time when the designer makes a statement that summarises how he or she *sees* the problem or, to be more specific, the structure of the situation that the problem presents.' They referred to this 'way of seeing the design situation' as the designer's 'problem paradigm'. As with their earlier studies of engineers, Lloyd and Scott found that the architects who had specific prior experience of the problem type had different approaches from their less-experienced colleagues: the experienced architects' approaches were characterised by strong problem paradigms, or 'guiding themes'. Cross and Clayburn Cross (98) have also identified, from interviews and protocol studies, the importance of problem framing, or the use of a strong guiding theme or principle, in the design behavior of outstanding, expert engineering designers. Darke (79) also reported from interviews with outstanding architects that they used strong guiding themes as 'primary generators' for setting problem boundaries and solution goals.

Schön (88) pointed out that 'the work of framing is seldom done in one burst at the beginning of a design process.' This was confirmed in Goel and Pirolli's (92) protocol studies of several types of designers (architects, engineers and instructional designers). They found that 'problem structuring' activities not only dominated at the beginning of the design task, but also reoccurred periodically throughout the task.

Valkenburg and Dorst (98) have attempted to develop and apply Schön's theory of reflective practice into team design activity, through a study of student industrial designers. In comparing a successful and an unsuccessful design team, Valkenburg and Dorst stressed the importance of the teams' problem framing. They identified five different frames used sequentially by the successful team during the project, in contrast to the single frame used by the unsuccessful team. The unsuccessful team also spent much greater amounts of time on 'naming' activities - i.e. on identifying potential problem features, rather than on developing solution concepts.

3 Solution Generation

The solution-focused nature of designer behavior appears to be appropriate behavior for responding to ill-defined problems. Such problems can perhaps never be converted to well-defined problems, and so designers quite reasonably adopt the more realistic strategy of finding a satisfactory solution, rather than expecting to be able to generate an optimum solution to a well-defined problem. However, this solution-focused behavior also seems to have potential drawbacks. One such drawback might be the 'fixation' effect induced by existing solutions.

3.1 Fixation

A 'fixation' effect in design was suggested by Jansson and Smith (91), who studied senior student and experienced professional mechanical engineers' solution responses to design problems. They compared groups of participants who were given a simple, written design brief, with those that were given the same brief but with the addition of an illustration of an existing solution to the set problem. They found that the latter groups appeared to be 'fixated' on the example design, producing solutions that contained many more features from the example design than did the solutions produced by the control groups. Jansson and Smith proposed that such fixation could hinder conceptual design if it prevents the designer from considering all of the relevant knowledge and experience that should be brought to bear on a problem. Some designers may be too ready to re-use features of known existing designs, rather than to explore the problem and generate new design features.

Purcell and Gero (91, 93, 96) undertook a series of experiments to verify and extend Jansson and Smith's findings on fixation. They studied and compared senior students in mechanical engineering and in industrial design. Early results suggested that mechanical engineers appeared to be much more susceptible to fixation than did industrial designers; the engineers' designs were substantially influenced by prior example designs, whereas the industrial designers appeared to be more fluent in producing a greater variety of designs, uninfluenced by examples. Purcell and Gero suggested that this might be a feature of the different educational programs of engineers and designers, with the latter being more encouraged to generate diverse design solutions. In a further development of the study, however, Purcell and Gero explored engineers' and designers' responses when the example design was an innovative rather than a routine prior solution. Here they found that engineers became fixated in the traditional sense when shown a routine solution, i.e. incorporating features of the routine solution in their own solutions, but became fixated on the underlying *principle* of the innovative solution, i.e. producing new, innovative designs embodying the same principle. The industrial designers, however, responded in similar ways under both conditions, generating wide varieties of designs that were not substantially influenced by any of the prior designs. Purcell and Gero therefore concluded that the industrial designers seem to be 'fixated on being different', and that 'fixation' in design may exist in a number of forms.

It is not clear that 'fixation' is necessarily a bad thing in design. As mentioned above, Cross and Clayburn Cross (98) have reported that outstanding engineering designers exhibit a form of 'fixation' on their problem frame, or on a guiding theme or principle. Having established the 'frame' for a particular problem, these designers can be tenacious in their pursuit of solution concepts that fit the frame. Similar observations have been reported by Lawson (94) and Darke (79) from their studies of outstanding architects. This tenacious fixation seems to be found often amongst highly creative individuals.

3.2 Attachment to concepts

Another form of 'fixation' that has been found to exist amongst designers is their attachment to early solution ideas and concepts. Although designers change goals and constraints as they design, they appear to hang on to their principal solution concept for as long as possible, even when detailed development of the scheme throws up unexpected difficulties and shortcomings in the solution concept. Some of the changing of goals and constraints during designing is associated with resolving such difficulties without having to start again with a major new concept. For example, from case studies of professional architectural design, Rowe (87) observed that: 'A dominant influence is exerted by initial design ideas on subsequent problem solving directions... Even when severe problems are encountered, a considerable effort is made to make the initial idea work, rather than to stand back and adopt a fresh point of departure.'

The same phenomenon was observed by Ullman et al. (88), in protocol studies of experienced mechanical engineering designers. They found that 'designers typically pursue only a single design proposal,' and that 'there were many cases where major problems had been identified in a proposal and yet the designer preferred to apply patches rather than to reject the proposal outright and develop a better one.' A similar observation was also made by Ball et al. (94), from their studies of senior students conducting 'real-world', final-year design projects in electronic engineering: 'When the designers were seen to generate a solution which soon proved less than satisfactory, they actually seemed loath to discard the solution and spend time and effort in the search for a better alternative. Indeed the subjects appeared to adhere religiously to their unsatisfactory solutions and tended to develop them laboriously by the production of various slightly improved versions until something workable was attained.'

Ball et al. regarded this behavior as indicating a 'fixation' on initial concepts, and a reliance on a simple 'satisficing' design strategy in contrast to any more 'well-motivated' process of optimisation. They found it difficult to account for this apparently unprincipled design behavior. Nevertheless, adherence to initial concepts and a satisficing strategy seem to be normal design behavior. Guindon (90a), in a study of experienced software designers, found that 'designers adopted a kernel solution very early in the session and did not elaborate any alternative solutions in depth. If designers retrieved alternative solutions for a subproblem, they quickly rejected all but one alternative by a trade-off analysis using a preferred evaluation criterion.'

However, in contrast to the 'fixation' findings reported above, in a study of senior industrial engineering students, Smith and Tjandra (98) found that the quality of design solutions produced did appear to be dependent upon a willingness to reconsider early concepts. They experimented with nine groups of four students, undertaking an artificial design exercise based upon two-dimensional configurations of coloured triangles supposed to have different functional properties. Each member of a design team in the exercise played a different rôle (architect, thermal engineer, structural engineer, and cost estimator). One of Smith and Tjandra's findings was that 'The top three designs . . . were the three groups that chose to scrap their initial design and to start afresh with a new design concept.' The successful players of this particular design game therefore seemed to be ones who were able and willing to overcome the possible fixation on an early concept. Perhaps it is worth emphasising that this study was based on a role-playing game and an artificial 'design' problem far removed from real-world design projects.

3.3 Generation of alternatives

It may be that good designers produce good early concepts that do not need to be altered radically during further development; or that good designers are able to modify their concepts rather fluently and easily as difficulties are encountered during development, without recourse to exploration of alternative concepts. Either way, it seems that designers are reluctant to abandon early concepts, and to generate ranges of alternatives. This does seem to be in conflict with a more 'principled' approach to design, as recommended by design theorists, and even to conflict with the idea that it is the exploration of solution concepts that assists the designer's problem understanding. Having more than one solution concept in play should promote a more comprehensive assessment and understanding of the problem.

Fricke (93, 96), from protocol studies of engineering designers, found that both generating few alternative concepts and generating a large number of alternatives were equally weak strategies, leading to poor design solutions. Where there was 'unreasonable restriction' of the search space (when only one or a very few alternative concepts were generated), designers became 'fixated' on concrete solutions too early. In the case of 'excessive expansion' of the search space (generating large numbers of alternative solution concepts), designers were then forced to spend time on organising and managing the set of variants, rather than on careful evaluation and modification of the alternatives. Fricke identified successful designers to be those operating a 'balanced search' for solution alternatives.

Fricke also found that the degree of precision in the problem as it was presented to the designers influenced the generation of alternative solution concepts. When the problem was precisely specified, designers generated more solution variants; with an imprecise assignment (for the same design task), designers tended to generate few alternative solution concepts. This perhaps indicates that the more active problem-framing required for an imprecise assignment leads more readily to preferred solution concepts. Designers given precise assignments have less scope for problem-framing, and generate a wider range of solution concepts in order to find a preferred concept.

3.4 Creativity

Designers themselves often emphasise the role of 'intuition' in the generation of solutions, and 'creativity' is widely regarded as an essential element in design thinking. Creative design is often characterised by the occurrence of a significant event, usually called the 'creative leap'. Recent studies of creative events in design have begun to shed more light on this previously mysterious (and often mystified) aspect of design.

Akin and Akin (96) studied creative problem solving behavior first on a classic problem where a form of 'fixation' normally prevents people from finding a solution to the problem: the 'nine-dots' problem. (In this problem, nine dots are arranged in a 3 x 3 square, and subjects are invited to join all nine dots by drawing just four straight lines without lifting pen from paper. Subjects normally assume that they have to draw within the implicit outline of the square, whereas the solution requires extending the lines to new vertices outside of the square.) They then extended their study from the nine-dot problem into a study of a simple architectural design problem, and compared the protocols of a non-architect and an experienced architect in tackling this problem. In these studies, Akin and Akin were looking for cases of the 'sudden mental insight' (SMI) that is commonly reported in cases of creative problem solving. They referred to the 'fixation' effect, such as the implicit nine-dot square, as a 'frame of reference' (FR) that has to be broken out of in order to generate creative alternatives. They suggested that a SMI occurs when a subject perceives their own fixation within a standard FR, and simultaneously perceives a new FR. The new FR also has to include procedures for generating a solution to the problem. The experienced architect possessed such procedural knowledge, whereas the novice did not, and was not able to generate anything other than a very conventional solution. Akin and Akin conclude: 'Realising a creative solution, by breaking out of a FR, depends on simultaneously specifying a new set of FRs that restructure the problem in such a way that the creative process is enhanced. The new FRs must, at a minimum, specify an appropriate *representational* medium (permitting the explorations needed to go beyond those of the earlier FRs), a design *goal* (one that goes beyond those achievable within the earlier FRs), and a set of *procedures* consistent with the representation domain and the goals.'

This seems to be similar to Schön's concept of a 'frame' which permits and encourages the designer to explore new design 'moves' and to reflect on the discoveries arising from those moves. But 'frames' can clearly be negative conceptual structures, when they are inappropriate 'fixations', as well as positive, creative structures.

Akin and Akin's conclusions also resonate with the study by Cross (97) of the 'creative event' that occurred in a protocol study of teamwork in industrial design. About half-way through the design session, one of the designers rather suddenly proposed that a concept for the device they were designing could be that 'maybe it's like a little vacuum-formed tray'. This was the first time in the team's activity that a 'tray' had been mentioned, but it immediately focused the team's attention and quickly became adopted as the basic concept for their solution. It appears to be the equivalent of a 'sudden mental insight', offering a new, creative 'frame of reference' meeting Akin and Akin's criteria, above. Cross suggested that the 'tray' concept should be seen not in terms of a 'creative leap', but in terms of a 'bridge' between the problem space and the solution space: 'During the design process, partial models of the problem and solution are constructed side-by-side . . . But the crucial factor, the "creative leap", is the bridging of these two partial models by the articulation of a concept (the "tray" idea in this example) which enables the models to be mapped onto each other. The "creative leap" is not so much a leap across the chasm between analysis and synthesis, as the throwing of a bridge across the chasm between problem and solution. The "bridge" recognisably embodies satisfactory relationships between problem and solution.'

It may be also that 'creative leaps' or 'sudden mental insights' are not so personal and idiosyncratic as has been promoted before. In protocol studies of experienced industrial designers, Cross and Dorst (98) observed that all nine subjects reported *the same* 'creative breakthrough'. All nine linked together the same pieces of available information and used this as a basis for their solution concept. All nine appeared to think that this was a unique personal insight.

3.5 The role of sketching

Several researchers have investigated the ways in which sketching helps to promote creativity in design thinking. Sketching helps the designer to find unintended consequences, the surprises that keep the design exploration going in what Schön and Wiggins (92) called the 'reflective conversation with the situation' that is characteristic of design thinking. Goldschmidt (91) called it the 'dialectics of sketching': a dialogue between 'seeing that' and 'seeing as', where 'seeing that' is reflective criticism and 'seeing as' is the analogical reasoning and reinterpretation of the sketch that provokes creativity. Goel (95) suggested that sketches help the designer to make not only 'vertical transformations' in the sequential development of a design concept, but also 'lateral transformations' within the solution space: the creative shift to new alternatives. Goel referred especially to the ambiguity inherent in sketches, and identified this as a positive feature of the sketch as a design tool.

It is not just formal or shape aspects of the design concept that are compiled by sketching; they also help the designer to identify and consider functional and other aspects of the design. Suwa, Purcell and Gero (98) suggested that sketching serves at least three purposes: as an external memory device in which to leave ideas as visual tokens, as a source of visuo-spatial cues for the association of functional issues, and as a physical setting in which design thoughts are constructed in a type of situated action. Although the above studies refer mostly to sketching in architectural design, Ullman et al. (90) also studied and emphasised the importance of sketching in mechanical engineering design, as have Kavakli et al. (98) and McGown et al. (98) in respect of product design. Verstijnen et al. (98) studied differences between skilled sketchers (industrial design students) and unskilled sketchers, and concluded that it was the skilled sketchers who benefited from the externalisation of mental imagery.

4 Process Strategy

An aspect of concern in design methodology and related areas of design research has been the many attempts at proposing systematic models of the design process, and suggestions for methodologies or structured approaches that should lead designers efficiently towards a good solution. However, most design in practice still appears to proceed in a rather ad-hoc and unsystematic way. Many designers remain wary of systematic procedures that, in general, still have to prove their value in design practice.

4.1 Structured process

It is not clear whether learning a structured, systematic process actually helps designers. One study that has suggested that a systematic approach might be helpful to students was that of Radcliffe and Lee (89). They studied senior students of mechanical engineering, working in small-groups (2-4) on a design project. In analysing the results, Radcliffe and Lee computed linear regression analyses of the subjects' design process sequence in comparison to an idealised, structured process of seven stages. They found that more 'efficient' processes (following closer to the supposed 'ideal') correlated positively with both quantity and quality of the subjects' design output: 'There was a positive correlation between the quality or effectiveness of a design and the degree to which the student follows a logical sequence of design processes.'

Fricke (93, 96) also studied a number of mechanical engineers, of varying degrees of experience and with varying exposures to education in systematic design processes. He found that designers following a 'flexible-methodical procedure' tended to produce good solutions. These designers worked reasonably efficiently and followed a fairly logical procedure, whether or not they had been educated in a systematic approach. In comparison, designers with too-rigid adherence to a methodical procedure (behaving 'unreasonably methodical'), or with very un-systematic approaches, produced mediocre or poor design solutions. It seems that, with or without an education in systematic design, designers need to exercise sophisticated strategic skills.

The occurrence of some relatively simple patterns of design process activity has often been suggested from anecdotal knowledge. For example, there has been a broad assumption that designing proceeds in cycles of analysis-synthesis-evaluation activities. Although such patterns of design process activity frequently have been proposed or hypothesised, there has been little empirical confirmation. McNeill et al. (98) were able to confirm some of these basic patterns, in a study of electronics engineers, using subjects with varying degrees of experience, from senior students to very experienced professionals. They were able to confirm that, 'In addition to the short-term cycles [of analysis-synthesis-evaluation], there is a trend over the whole design episode to begin by spending most of the time analysing the problem, then mainly synthesising the solution and finishing by spending most time on the evaluation of the solution.' They also confirmed a supposed progression through the design process from first considering required functions, then structure of potential solutions, and then the behavior of those solutions. Their general, if unsurprising conclusion was that: 'A designer begins a conceptual design session by analysing the functional aspects of the problem. As the session progresses, the designer focuses on the three aspects of function, behavior and structure, and engages in a cycle of analysis, synthesis and evaluation. Towards the end of the design session, the designer's activity is focused on synthesising structure and evaluating the structure's behavior.'

4.2 Opportunism

In contrast to studies that confirm the prevalence and relevance of fairly structured design behavior, there have also been some reports from studies that have emphasised the 'opportunistic' behavior of designers. This emphasis has been on designers' deviations from a structured plan or methodical process into the 'opportunistic' pursuit of issues or partial solutions that catch the designer's attention. For example, Visser (90) made a longitudinal study of an experienced mechanical engineer, preparing a design specification. The engineer claimed to be following a structured approach, but Visser found frequent deviations from this plan. 'The engineer had a hierarchically structured plan for his activity, but he used it in an opportunistic way. He used it only as long as it was profitable from the point of view of cognitive cost. If more economical cognitive actions arose, he abandoned it.' Thus Visser regarded reducing 'cognitive cost' - i.e. the cognitive load of maintaining a principled, structured approach - as a major reason for abandoning planned actions and instead delving into, for example, confirming a partial solution at a relatively early stage of the process.

From protocol studies of three experienced software system designers, Guindon (90b) also emphasised the 'opportunistic' nature of design activities. Guindon stressed that 'designers frequently deviate from a top-down approach. These results cannot be accounted for by a model of the design process where problem specification and understanding precedes solution development and where the design solution is elaborated at successively greater levels of detail in a top-down manner.' Guindon observed the interleaving of problem specification with solution development, 'drifting' through partial solution development, and jumps into exploring suddenly-recognised partial solutions, which were categorised as major causes of 'opportunistic solution development'. Guindon also referred to 'cognitive cost' as one possible explanation for such behavior: 'Designers find it advantageous to follow a train of thought temporarily, thus arriving at partial solutions at little cognitive cost.'

Ball and Ormerod (95) criticised a too-eager willingness to emphasise 'opportunism' in design activity. In their studies of expert electronics engineers they found very few deviations from a top-down, breadth-first design strategy. But they did find some significant deviations occurring, when designers made a rapid depth-first exploration of a solution concept in order to assess its viability. Ball and Ormerod did not regard such occasional depth-first explorations as implying the abandonment of a structured approach. Instead, they suggested that expert designers will normally use a mixture of breadth-first and depth-first approaches: 'Much of what has been described as opportunistic behavior sits comfortably within a structured top-down design framework in which designers alternate between breadth-first and depth-first modes.' Ball and Ormerod were concerned that 'opportunism' seemed to imply unprincipled design behavior, 'a non-systematic and heterarchical process' in contrast to the assumed ideal of a systematic and hierarchical process. However, rather than regarding opportunism as unprincipled design behavior, Guindon had suggested it might be inevitable in design: 'These deviations are not special cases due to bad design habits or performance breakdowns but are, rather, a natural consequence of the ill-structuredness of problems in the early stages of design.'

4.3 Modal shifts

An aspect of cognitive strategy that emerges from several studies is that, especially during creative periods of conceptual design, designers alternate rapidly in shifts of attention between different aspects of their task, or between different modes of activity. Akin and Lin (95), in their protocol study of an experienced engineering designer, first identified the occurrence of 'novel design decisions' (NDDs). These, in contrast to routine design decisions, are decisions that are critical to the development of the design concept. Akin and Lin also segmented the designer's activities into three modes: drawing, examining and thinking. Then, allowing for some implicit overlap or carry-over of the designer's attention from one segment to another, they represented the designer's activities in terms of single-, dual- or triple-mode periods. They found a significant correlation between the triple-mode periods and the occurrence of the NDDs: 'Six out of a total of eight times a novel design decision was made, we found the subject alternating between these three activity modes (examining-drawing-thinking) in rapid succession.' Akin and Lin are cautious about drawing any inference of causality, concluding only that 'Our data suggest that designers explore their domain of ideas in a variety of activity modes . . . when they go beyond routine decisions and achieve design breakthroughs.'

Some studies of student designers have also noted the apparent importance of frequent shifts of attention or activity mode in influencing either the creativity or overall quality of the design concepts produced. For example, in their protocol studies of junior and senior students of industrial design, Cross et al. (94) segmented the students' activities into the three modes of gathering information, sketching and reflecting. They suggested that the more successful students (in producing creative design concepts) were those who showed evidence of rapid alternation between the activity modes. Also, Atman et al. (99), from their study of freshmen and senior engineering design students, suggested that overall quality of design concepts was related to rapid alternation of activities, which they measured as transitions between design steps such as gathering information, generating ideas and modelling.

4.4 Novices and experts

Novice behavior is usually associated with a 'depth-first' approach to problem solving, i.e. sequentially identifying and exploring sub-solutions in depth, whereas the strategies of experts are usually regarded as being predominantly top-down and breadth-first approaches. But this may be too simplistic a view of the reality of process strategy in design. Ball and Ormerod's (95) comments about top-down, structured approaches *vs.* 'opportunism' have been noted above. They concluded that 'it would be surprising if it is practicable for expert designers to adopt a purely breadth-first or depth-first approach. Indeed, a flexible mixture of modes is a more psychologically realistic control structure for expert design.' They suggested that, whilst a depth-first approach minimises cognitive load, a breadth-first approach minimises commitment and optimises design time and effort. Those suggestions would also quite reasonably reflect the respective concerns and strategies that we might expect of novices and experts.

Many of the classic studies of expertise have been based on examples of game-playing (e.g. chess), or on comparisons of experts versus novices in solving routine problems (e.g. physics). These are all well-defined problems, whereas designers characteristically deal with ill-defined problems. Some studies of expertise in fields such as creative writing and computer programming (Holyoak (91); Adelson and Solway (88)) where problems are more ill-defined, do suggest some parallels with observations of expert designers. These studies suggest that some of the 'standard' results from studies of expertise do not match with results from studies of expertise in creative domains. For example, creative experts will define the given task so that it is problematic - i.e. deliberately treat it as ill-defined - which is contrary to the assumption that experts will generally solve a problem in the 'easiest' way, or certainly with more ease than novices. In some ways, therefore, creative experts treat problems as 'harder' problems than novices do. Creative experts are also reported as solving similar tasks from first principles each time, rather than recalling previous solutions. Cross and Clayburn Cross (98) have suggested that such features of creative expertise can be seen in the strategies of outstanding expert designers.

Göker (97) compared novices and experts performing design-related problem solving tasks the computer-simulated construction of 'machines' from catalogues of parts to achieve certain objectives. A particularly unusual aspect of this study was the experimental method, based on the use of electro-encephalograph (EEG) records. Göker found that experts (subjects skilled in the use of the computer simulation) used more of the visuo-spatial regions of their brains than did the novices, who used more of their brain regions associated with verbal-abstract reasoning. The implication is that experts do not 'reason' towards a design concept in an abstract way, but rely more on their experience and on visual information.

5 Conclusions

In this chapter I have surveyed the growth of empirical studies of design cognition, and attempted to draw out some of the issues that have emerged in such studies. I have taken a cross-disciplinary view, and looked for comparisons across the different domains of professional design practice. There has been a number of striking similarities identified in design activity, independent of professional domain, suggesting that design cognition is indeed a domain-independent phenomenon.

I have concentrated on protocol and similar formalized methods of study, and I have therefore omitted a wide range of other kinds of studies that also have relevant and important contributions to make to the understanding of design cognition and the nature of design activity. Protocol analysis has some severe limitations as a research method for investigating design activity - for instance, it is extremely weak in capturing non-verbal thought processes, which are so important in design work (Lloyd et al. 95). Dorst and Cross (97) concluded from the Delft Design Protocols Workshop that protocol analysis provides a very valuable but highly specific research technique, capturing a few aspects of design thinking in detail, but failing to encompass many of the broader realities of design in context. Other kinds of study, which attempt to capture a broader view, include detailed observation of industrial practice, such as Frankenberger and Badke-Schaub (98), and ethnographic methods, such as Bucciarelli (94). There has also been valuable historical work, such as Ferguson's (92) study of the role of drawing in engineering, and significant theoretical contributions to identifying fundamental aspects of design reasoning and logic, such as Roozenburg (93).

The range and number of studies surveyed in this paper suggest that the field of empirical studies of design activity is continuing to grow, and a number of shared issues has been identified. In many cases, these issues remain unresolved, and there is therefore still considerable work to be done to establish a robust and reliable understanding of design cognition.

5.1 Summary: problem formulation

• *Goal analysis*

Designers appear to be 'ill-behaved' problem solvers, in that they do not spend much time and attention on defining the problem. However, this seems to be appropriate behavior, since some studies have suggested that over-concentration on problem definition does not lead to successful design outcomes. It appears that successful design behavior is based not on extensive problem analysis, but on adequate 'problem scoping' and on a focused or directed approach to gathering problem information and prioritising criteria. Setting and changing goals are inherent elements of design activity.

• *Solution focusing*
Designers are solution-focused, not problem-focused. This appears to be a feature of design cognition which comes with education and experience in designing. In particular, experience in a specific problem domain enables designers to move quickly to identifying a problem 'frame' and proposing a solution conjecture

• *Problem/solution co-evolution*
The concept of 'co-evolution' of both the problem and its solution has been proposed to describe how designers develop both aspects together in conceptual stages of the design process. The designer's attention oscillates between the two, forming partial structurings of the two 'spaces' of problem and solution. Designing appears to be an 'appositional' search for a matching problem-solution pair, rather than a propositional argument from problem to solution.

• *Problem framing*
Processes of structuring and formulating the problem are frequently identified as key features of design activity. The concept of 'problem framing' perhaps seems to capture best the nature of this activity. Successful, experienced and - especially - outstanding designers are repeatedly found in various studies to be pro-active in problem framing, actively imposing their view of the problem and directing the search for solution conjectures.

5.2 Summary: solution generation

• *Fixation*
'Fixation' seems to be double-edged feature of design activity, in that it can lead to conservative, routine design or - perhaps only when exercised by outstanding designers - to creative, innovative design. There may be differences between educational programs of engineers and industrial designers (and probably architects) which lead engineers more readily to fixate on features of prior design solutions.

• *Attachment to concepts*
Designers become readily attached to single, early solution concepts and are reluctant to abandon them in the face of difficulties in developing these concepts into satisfactory solutions. This seems to be a weak feature of design behavior, which may be susceptible to change through education. However, trying to change the 'unprincipled' and 'ill-behaved' nature of conventional design activity may be working against aspects that are actually effective and productive features of intuitive design cognition.

• *Generation of alternatives*
Generating a wide range of alternative solution concepts is another aspect of design behavior which is recommended by theorists and educationists but appears not to be normal design practice. Generating a very wide range of alternatives may not be a good thing: some studies have suggested that a relatively limited amount of generation of alternatives may be the most appropriate strategy.

• *Creativity*
Creative thinking has tended to be regarded as mysterious, but new explanatory descriptions of creativity in design are beginning to emerge from empirical studies. In particular, it no longer seems correct to promote the key feature of creative design as dependent upon an intuitive, heroic 'creative leap' from problem to solution. Problem framing, co-evolution, and conceptual bridging between problem space and solution space seem to be better descriptors of what actually happens in creative design.

• *Sketching*
The key 'tool' to assist design cognition remains the traditional sketch. It seems to support and facilitate the uncertain, ambiguous and exploratory nature of conceptual design activity. Sketching is tied-in very closely with features of design cognition such as the generation and exploration of tentative solution concepts, the identification of what needs to be known about the developing concept, and especially the recognition of emergent features and properties. Studies of the role of sketching have all emphasised its inherent power as a design aid.

5.3 Summary: process strategy

• *Structured process*
Following a reasonably-structured process seems to lead to greater design success. However, rigid, over-structured approaches do not appear to be successful. They key seems to be flexibilty of approach, which comes from a rather sophisticated understanding of process strategy and its control.

• *Opportunism*
'Opportunistic' behavior sounds like another feature of the characteristically 'unprincipled', 'ill-behaved' activity of designers. As with some other aspects of intuitive design behavior, it may be that we should not equate 'opportunistic' with 'unprincipled' behavior in design, but rather that we should regard 'opportunism' as characteristic of expert design behavior. The 'cognitive cost' of apparently more principled, structured behavior may actually be higher than can be reasonably sustained, or can be justified by quality of outcome.

• *Modal shifts*

It has been noticed in some studies that creative, productive design behavior seems to be associated with frequent switching of types of cognitive activity. There is no clear explanation for this observation, but it may be related to the need to make rapid explorations of problem and solution in tandem.

• *Experts and novices*

Conventional wisdom about the nature of problem solving expertise seems often to be contradicted by the behavior of expert designers. In design education we must therefore be very wary about importing models of behavior from other fields. Empirical studies of design activity have frequently found 'intuitive' features of design behavior to be the most effective and relevant to the intrinsic nature of design. Some aspects of design theory, however, have tried to develop counter-intuitive models and prescriptions for design behavior. We still need a much better understanding of what constitutes expertise in design, and how we might assist novice students to gain that expertise.

References

Adelson B, and Solway E, (1985), The role of domain experience in software design, *IEEE Transactions on Software Engineering,* 11, 1351-1360.

Adelson B, and Solway E, (1988), A model of software design, M Chi R Glaser and M Farr Ed. *The nature of expertise*, Lawrence Erlbaum and Associates, Hillsdale, NJ.

Akin Ö, (1978), How do architects design? J C Latombe Ed. *Artificial intelligence and pattern recognition in computer aided design,* IFIP/North-Holland, New York, NY.

Akin Ö, and Akin C, (1996), Frames of reference in architectural design: Analysing the hyper-acclamation (Aha!), *Design Studies,* 17(4), 341-361.

Akin Ö, and Lin C, (1995), Design protocol data and novel design decisions, *Design Studies,* 16(2), 211-236.

Atman C, Chimka J, et al. (1999), A comparison of freshman and senior engineering design processes, *Design Studies,* 20(2), 131-152.

Ball L J, Evans J, et al. (1994), Cognitive processes in engineering design: A longitudinal study, *Ergonomics,* 37(11): 1753-1786.

Ball L J, and Ormerod T, (1995), Structured and opportunistic processing in design: a critical discussion, *International Journal of Human-Computer Studies,* 43, 131-151.

Ballay J, (1987), An experimental view of the design process, W B Rouse and K B Roff Eds. *System design: Behavioral perspectives on designers, tools and organisations,* North-Holland, New York, NY.

Bucciarelli L, (1994), *Designing engineers*, MIT Press, Cambridge, MA.

Chan C-S (1990), Cognitive processes in architectural design problem solving, *Design Studies* 11(2), 60-80.

Christiaans H, and Dorst C. (1992), Cognitive models in industrial design engineering: A protocol study, D L Taylor and D A Stauffer Eds. *Design theory and methodology - DTM92,* American Society of Mechanical Engineers, New York, NY.

Colgan L and Spence R, (1991), Cognitive modelling of electronic design, J Gero Ed. *AI in design 91,* Butterworth-Heinemann, Oxford, UK.

Cross N, (1997), Creativity in design: Analyzing and modeling the creative leap, *Leonardo,* 30(4), 311-317.

Cross N, (1999), Natural intelligence in design, *Design Studies,* 20(1), 25-39.

Cross N, Christiaans H, and Dorst K, (1994), Design expertise amongst student designers, *Journal of Art and Design Education,* 13(1), 39-56.

Cross N, Christiaans H, and Dorst K, Ed. (1996), *Analysing design activity*, John Wiley & Sons, Chichester, UK.

Cross N, and Clayburn Cross A, (1998), Expertise in engineering design, *Research in Engineering Design,* 10(3), 141-149.

Cross N, and Dorst K, (1998), Co-evolution of problem and solution spaces in creative design: observations from an empirical study, J Gero and M L Maher Eds. *Computational models of creative design IV,* University of Sydney, NSW, Australia.

Darke J, (1979), The primary generator and the design process, *Design Studies,* 1(1), 36-44.

Davies S, and Castell A, (1992), Contextualizing design: Narratives and rationalization in empirical studies of software design, *Design Studies,* 13(4), 379-392.

Dorst K, and Dijkhuis J, (1995), Comparing paradigms for describing design activity, *Design Studies,* 16(2), 261-274.

Eastman C M, (1969), Cognitive processes and ill-defined problems: A case study from design, *International joint conference on artificial intelligence*, Washington, DC.

Eastman C M, (1970), On the analysis of intuitive design processes, G T Moore Ed. *Emerging methods in environmental design and planning,* MIT Press, Cambridge, MA.

Eckersley M, (1988), The form of design processes: A protocol analysis study, *Design Studies,* 9(2), 86-94.

Ehrlenspiel K, Dylla N, et al. (1992), Experimental investigation of individual processes in engineering design, N Cross K Dorst and N Roozenburg Ed. *Research in design thinking,* Delft University Press, Delft, The Netherlands.

Ferguson E, (1992), *Engineering and the mind's eye*, MIT Press, Cambridge, MA.

Foz A, (1973), Observations on designer behavior in the *parti DMG-DRS journal: Design Reseach and Methods,* 7(4), 320-323.

Frankenberger E, and Badke-Schaub P, (1998), Integration of group, individual and external influences in the design process, E Frankenberger P Badke-Schaub and H Birkhofer Eds. *Designers - the key to successful product development,* Springer, London, UK.

Fricke G, (1993), Empirical investigations of successful approaches when dealing with differently précised design problems, *International conference on engineering design ICED93*, Heurista, Zürich.

Fricke G, (1996), Successful individual approaches in engineering design, *Research in Engineering Design,* 8, 151-165.

Goel V, (1995), *Sketches of thought*, MIT Press, Cambridge, MA.

Goel V, and Pirolli P, (1992), The structure of design problem spaces, *Cognitive Science,* 16, 395-429.

Göker M H, (1997), The effects of experience during design problem solving, *Design Studies,* 18(4), 405-426.

Goldschmidt G, (1991), The dialectics of sketching, *Creativity Research,* 4(2), 123-143.

Guindon R, (1990a), Knowledge exploited by experts during software system design *I J Man-Machine Studies,* 33, 279 304.

Guindon R, (1990b), Designing the design process: Exploiting opportunistic thoughts, *Human-Computer Interaction,* 5, 305-344.

Holyoak K J, (1991), Symbolic connectionism: Toward third-generation theories of expertise,, K A Ericsson and J Smith Eds. *Toward a general theory of expertise: Prospects and limits,* Cambridge University Press, Cambridge, UK.

Jansson D G, and Smith S M, (1991), Design fixation, *Design Studies,* 12(1), 3-11.

Jeffries R, Turner A, et al. (1981), The processes involved in designing software, J R Anderson Ed. *Cognitive skills and their acquisition*, Lawrence Erlbaum and Associates, Hillsdale, NJ.

Kavakli M, and Scrivener S, et al. (1998), Structure in idea sketching behavior, *Design Studies,* 19(4), 485-517.

Kolodner J L, and Wills L M, (1996), Powers of observation in creative design, *Design Studies,* 17(4), 385-416.

Lawson B, (1979), Cognitive strategies in architectural design, *Ergonomics,* 22(1), 59-68.

Lawson B, (1994), *Design in mind*, Butterworth-Heinemann, Oxford, UK.

Lloyd P, Lawson B, and Scott P, (1996), Can concurrent verbalisation reveal design cognition? *Design Studies,* 16(2), 237-259.

Lloyd P, and Scott P, (1994), Discovering the design problem, *Design Studies,* 15(2), 125-140.

Lloyd P, and Scott P, (1995), Difference in similarity: Interpreting the architectural design process, *Planning and Design,* 22, 383-406.

McGown A, Green G, et al. (1998), Visible ideas: Information patterns of conceptual sketch activity, *Design Studies,* 19(4), 431-453.

McNeill T, Gero J, et al. (1998), Understanding conceptual electronic design using protocol analysis, *Research in Engineering Design,* 10(3), 129-140.

Olson G M, and Olson J S, et al. (1992), Small group design meetings: An analysis of collaboration, *Human-Computer Interaction,* 7, 347-374.

Purcell A T, and Gero J, (1991), The effects of examples on the results of design activity, J S Gero Ed. *Artificial intelligence in design AID91,* Butterworth-Heinemann, Oxford, UK.

Purcell T, and Gero J, (1996), Design and other types of fixation, *Design Studies,* 17(4), 363-383.

Purcell A T, Williams P, et al. (1993), Fixation effects: Do they exist in design problem solving? *Environment and Planning B: Planning and Design,* 20(3), 333-345.

Radcliffe D, and Lee T Y, (1989), Design methods used by undergraduate engineering students, *Design Studies,* 10(4), 199-207.

Roozenburg N, (1993), On the pattern of reasoning in innovative design, *Design Studies,* 14(1), 4-18.

Rowe P, (1987), *Design thinking*, MIT Press, Cambridge, MA.

Schön D A, (1983), *The reflective practitioner*, Temple-Smith, London, UK.

Schön D A, (1988), Designing: Rules, types and worlds, *Design Studies,* 9(3), 181-190.

Schön D, and Wiggins G, (1992), Kinds of seeing and their functions in designing, *Design Studies,* 13(2), 135-156.

Smith R P and Tjandra P, (1998), Experimental observation of iteration in engineering design, *Research in Engineering Design,* 10(2), 107-117.

Stauffer L, and Ullman D, et al. (1987), Protocol analysis of mechanical engineering design, *International conference on engineering design ICED87*, ASME, New York, NY.

Suwa M, Purcell T, and Gero J, (1998), Macroscopic analysis of design processes based on a scheme for coding designers' cognitive actions, *Design Studies,* 19(4), 455-483.

Suwa M, and Tversky B, (1997), What do architects and students perceive in their design sketches? *Design Studies,* 18(4), 385-403.

Tang J C, and Leifer L, (1990), An observational methodology for studying group design activity, *Research in Engineering Design,* 2, 209-219.

Thomas J C, and Carroll J M, (1979), The psychological study of design, *Design Studies* 1(1), 5-11.

Ullman D G, and Dietterich T G, et al. (1988), A model of the mechanical design process based on empirical data, *AI in Engineering Design and Manufacturing,* 2(1), 33-52.

Ullman D G, and Wood S, et al. (1990), The importance of drawing in the mechanical design process, *Computers and Graphics,* 14(2), 263-274.

Valkenburg R, and Dorst K, (1998), The reflective practice of design teams, *Design Studies,* 19(3), 249-271.

Verstijnen I M, Hennessey J M, Leeuwen C van, Hamel R, and Goldschmidt G, (1998), Sketching and creative discovery, *Design Studies*, 19 (4), 519-546.

Visser W, (1990), More or less following a plan during design: Opportunistic deviations in specification, *International Journal of Man-Machine Studies,* 33, 247-278.

Chapter 6
Variants in Design Cognition

Ömer Akin
School of Architecture
Carnegie Mellon University
Pittsburgh, PA

Abstract

In different fields of design, cognitive processes have both similarities and differences. Some of the notable differences that distinguish architectural design from other design fields are rich representations, indiscriminate use of inventive strategies, non-standard problem composition schema, and strategies of complexity management.

1 How Design Disciplines Differ

This chapter is about cognitive models of designers and how these models depict different design professions. Before I present some of the empirical bases for making claims in this area, let us start with some popular notions about professional differences. For some, these notions set the stage for the more serious explorations of professional differences and similarities.

1.1 Architects and Engineers

A well-known and sometimes overused joke about the profession goes something like this. Architects begin by knowing a little about a lot of things and end up knowing nothing about everything; conversely, engineers start by knowing a lot about a few things and end up knowing everything about nothing. Most see the humor in the joke and some appreciate a degree of truth in it. At the risk of abandoning the humor I would like to talk about another distinction which should get us closer to cognitive issues.

I claim that the profession of architecture rewards the heart while engineering rewards the brain. Architects who win awards and reach the highest levels of the profession do so primarily because and in spite of the foolish risks they have taken. Some examples would include the Sydney Opera House, Centre Georges Pompidou, and the Fallingwater. In all of these cases architects have ventured into the unknown territory of structural design, systems design and weather protection and by some accounts have failed miserably. Yet their works are considered masterpieces from which countless generations of young architects learn that all but unreachable glory of creating another Guggenheim Museum or Ronchamp Church is worth overriding the tenets of rationality in their work.

Engineers on the other hand have been condemned for the heartless mistakes they have made in service of a blind trust in technology; as in the cases of the Hartford Civic Center, the Tacoma Narrows Bridge, and the Kansas City Hyatt. In these cases, engineers were shielded by their professional authority from questioning and being questioned about their designs in spite of concerns raised by field observations (Akin 99). Somehow the results obtained through objective methods, however abstract and limited in scope, were allowed to override the tragic results foretold by seemingly spurious evidence.

Due to my work with architects, I looked and am confident that architects do have brains (let's reintroduce a modest amount of humor). Likewise it is easy to imagine that engineers have hearts, as well. It is a matter of re-thinking our priorities in architecture so that we have architects acting after they filter with their brains what their hearts have been telling them. In this paper, I will try to describe what the brain (rather the cognitive system) of the architect is able to do and is urged to do by the domain of problems with which she is accustomed to deal. This is not to say that there are biological or physiological differences, or even fundamental intelligence related differences between architects and others. I believe it is a matter of ethos and culture fostered in a given profession, its educational philosophy and the predisposition of its participants.

1.2 Architects and Scientists

Before moving into the behavioral variants and invariants that distinguish the cognition of architects, let me indulge in one more comparison. Architects, this time as opposed to scientists, stake their claim on the basis of *situated persuasiveness*. They argue not that their solutions are absolutely inescapable but that they are the most supportable ones under the circumstances of given problems. Change any assumption, the solution will change. Even under the same set of assumptions, there is no assurance that the selected solution is the best one since the resources allocated to architectural projects are insufficient to exhaust all possible solutions. Therefore the architect has to make her stand on the basis of persuading others (clients, owners, users, public officials, consultants, financiers, contractors, and developers). The solution they foresee merely has the promise of satisfying the requirements of the problem (Simon 73).

This is far different from the stand of the scientist or the engineer, for that matter, who must make her argument on the basis that all other conceivable possibilities have been refuted or an objective measurement of the *best choice* can be made by virtue of optimization. Karl Popper (63) articulates the scientist's mode of operation in his theory of *Falsificationism*. The only way of knowing if a scientific theory would hold is to put it to the most stringent tests. This requires that the scientist sincerely and vigorously become an antagonist to her own theories. By being the severest critic of them she has a chance of eliminating the worst criticism that can be thrown at them. This is a quest of search for the truth, the absolute truth. The truths that architects seek are temporal ones that are situated in a specific context.

2 Variants and Invariants

In some respects the field of Cognitive Psychology and even the entire domain of empirical research is about recognizing predictability of measurable events. One way of achieving this is to find the conditional invariants among all of the variables that are of interest. The conditions of such an inquiry are represented by the *independent* variables. To the extent that one makes good assumptions about the relationships between these and the *dependent* variables, not to mention the *uncontrolled* ones (Figure 1); finding dependent variables whose values are *invariant* makes any scientist's day, often career.

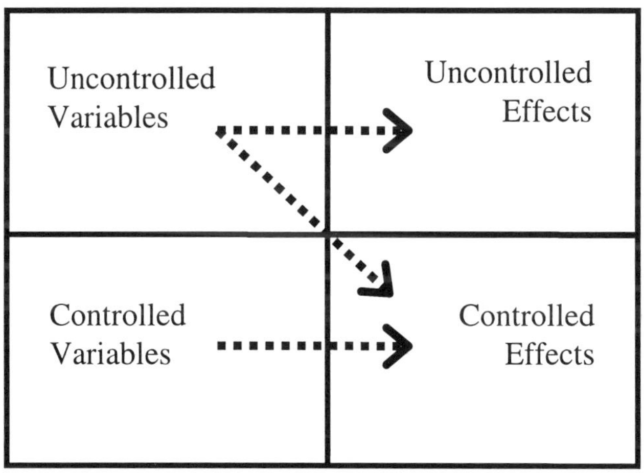

Figure 1: Relationship of controlled-uncontrolled variables and effects (Levin, 1984)

Discriminating invariants or, for that matter, the variants (as defined above) should then be the passion of the researcher interested in design cognition. What do architects do that others do not under similar conditions? Given that we all have the same brainware, give or take a few synapses, what cognitive activities remain constant across tasks and individuals? We can throw in a few more questions in the area of design education since this is a related agenda item. Given that we instruct students about design in many different fields, what justification is there to have domain-specific design courses; or is there any justification to have a common core for different design disciplines?

In order to answer these questions, this paper will cull a handful of salient cognitive features of the architectural design activity, from a collection of seminal studies on design cognition. I will rely on others work as well as mine but will draw primarily from my research experiences in order to formulate a more coherent viewpoint.

This neither says that those characteristics, however coherently stated, constitute a sufficient model of design nor that they are mutually exclusive with other fields. In order to claim the former, one would have to concentrate exclusively on the invariants of design. In order to claim the latter, one would have to be an expert in all design fields, which is what I am not.

2.3 Invariants of Architect's Cognition

For nearly two decades I worked on describing the invariants across individual differences, tasks, and domains, in architectural design. The individual differences dealt with those between experts and novices (Akin 86a). I observed that experts possess larger chunks (as predicted by performance in other domains such as chess, Chase and Simon 73). Experts also use a hybrid search strategy of breadth-first-depth-next. When experts search through alternatives they generate many more new problem structures, or formulations, than non-experts (Akin 86b). This result holds even in cases where an early solution is found and there still is time to design. In dealing with different cognitive task domains, I observed that, there are notable differences between puzzles, decision problems, and design (Akin 97). Design, due to the ill-defined nature of the task, elicits spurious problem re-structurings, and frequent shifts between representations. I also observed that, in problem solving and puzzles, restructuring is less frequent but strategic in leading to a solution. The explicit relationship of a newly defined representation to the solution set is often a prerequisite for success.

There are many more lessons learned from these studies. It would be a straightforward matter to cite them here. However, the more interesting task, which I have decided to give myself, is to describe the *variants*: those cognitive skills that characterize architecture but not other fields. Through this, I intend to implicitly define the range for the opposite category: the invariants.

Studies of architectural cognition concern themselves with a few overarching issues: *representation*, *strategic behavior*, and *innovation*. These in turn have many other sub-issues. For example representation can be decomposed into sketching and analog representation, knowledge representation, problem representation, and so on. Strategic behavior subsumes issues like search, problem restructuring, process management, and so on. The last category, innovation, can be seen as a subset of strategic behavior as well.

However, there is a growing field of inquiry on design creativity, which fits neither into one or the other of the previous two categories.

Surveying these principal categories of design cognition (representation, strategic behavior and innovation) I have identified four, more specific cognitive behaviors that I regard as principal variants of the field of architecture. These consist of (1) rich representations, (2) indiscriminate use of creative design strategies, (3) non-standard problem composition schemata, and (4) complexity management approaches. Each of these is, admittedly, related to representation, strategic behavior and creation, however, their scope in the design process is all but invariant. Thus, in the following sections, I will primarily discuss these four topics.

2.4 Variants of the Architect's Cognition

2.2.1 Analog versus Symbolic Representation

In the area of problem solving the argument about the value of appropriate representations has been made more powerfully than in most other areas. When isomorphic representations are used to present identical problems, differences in problem understanding and problem solving have been observed (*Tower of Hanoi* versus *Tea Ceremony,* Newell and Simon 72; *Mutilated Checkerboard*, Kaplan 90). Some representations are better suited to solve certain problems. The Tea Ceremony problem, for example, which is an isomorph of the Tower-of-Hanoi problem, appears to be more difficult to solve than the latter. This is largely due to its presentation (or representation) in the narrative form as opposed to the colored pegs and discs used in solving the Tower-of-Hanoi problem.

Problem representation is a critical tool that influences problem solving independently of field or discipline. de Kleer, in studying how physics problems are solved, for example, differentiates between domain knowledge, which can be *physical* or mathematical, and *envisionment* (75, 77), which is domain independent. More recently, Larkin (83) described these as *physical* and *naïve* representations, respectively. Envisionment, or naïve representations, depicts problems of Physics through everyday objects in relation to one another (pulleys and ropes, rollers and inclines). Thus the ordinary problem solver, or by analogy the "naïve" problem solver, has to find suitable operations that work in this representational domain. These include things like motion, collision, and friction and help the problem solver envision what could happen in the context of the given problem.

The expert, on the other hand, describes the same situation through *physical* entities, such as forces, energy and the like. She sets up mathematical relationship between these entities, such as FORCE = MASS*GRAVITY*HEIGHT. The proper solution of almost all such problems may initially use *naïve* representations but ultimately depend on the construction of accurate *physical* representations.

In the architectural domain, an analogous condition can be seen in the example of the stacking and blocking problem. Given a problem of clustering of building functions into logical groups and sub-groups, the nodes of a bubble-diagram can be extremely useful in creating a (naïve) representation of these functions. In comparison, a matrix of numerical values accurately indicate the spatial distance between pairs of functions. The twist in this comparison is that the matrix format is not the most suitable one for the designer. Most designers prefer to work with bubble diagrams due to the ease of direct manipulation of functions and their relationships. Alternatively, the matrix representation is preferred over the former, in solving the problem algorithmically. Thus, while we can identify the equivalents to naïve and physical representations in design, their relation to domain knowledge is not as obvious.

Architects and other designers working with graphic media deal with multiple representations. Architects primarily rely on the equivalent of the *naïve* representation as a primary environment of work. They depict everyday objects in space through drawings, sketches, physical and electronic models. This, however, is anything but domain independent or "naïve" in the true sense of that word. What appears to be the everyday realm, in this case, is in fact the specialty domain of the architect. Consequently, I will call this category of representations, *analog*, in order to make the name of the category value neutral. In contradistinction, physical and mathematical representations will be called *symbolic* representations, since they abstract the real world into symbols with or without their topological relations consistently maintained.

The power of *analog* representations used by the architect is based on the directness of their correspondence to reality, the accuracy with which they simulate objects and the evaluation of important design performance issues they enable, such as composition, contextual congruency, and constructability. This is why there has been such an emphasis on sketching, in architectural education and practice. Researchers interested in describing this process and building computer systems around them have proposed models that describe the role of analog representations in design. Goel (92), for example, describes the logic behind drawings that are used in ill-structured problems such as design. The ambiguous nature of these representations, in fact, helps in the process of exploring ill-structured problems. Goldschmidt (91) speaks directly about the symbiotic relationship between internal (cognitive) and external (analog) representations and shows how the give and take between them is critical for pushing design forward. Finally, Larkin *et.al.* (87) show the importance of analog representations even for problem solving domains outside of design.

In design, *symbolic* representations are also important. Designers are not only interested in performance issues that can be simulated through analog or topologically correct representations. They also want to know about heat transfer, light and sound distribution, and transfer of loads and moments through materials. All of these phenomena require the kind of abstraction found in Physics problems. Consequently, the design process has to integrate symbolic and analog representations in order to solve multi-constraint design problems.

2.2.2 Dimensions of Architectural Representation

This picture is not limited to the domain of architecture. All design domains rely on a relatively rich set of representations. Industrial designers, mechanical engineers, and electrical engineers alike use *symbolic* as well as *analog* representations. They explore possibilities, conduct analyses, and evaluate potential solutions before arriving at a final design. There is ample domain-specific evidence to support the need for representations particularly suited for specific design tasks, that is, if the resulting design is going to be successful.

Some of these representations are uniformly used in all stages of the design process. For example technical specifications of an aircraft engine are given at the outset as part of the performance requirements and are tracked throughout the design process as constraints to be satisfied by different design approaches or as evaluation criteria. Other representations may come into the picture on an as needed basis. For example, airflow diagrams in wind tunnel testing or the simulation of aerodynamic behavior would be introduced at a stage when the three-dimensional form of the engine is being designed.

In architectural design, the range and scope of representations used during different stages of the process are broader than they are in other design domains. This is one of the significant differences between architecture and other design areas, particularly those outside of *situated problem solving* domains like urban design, site design and structural engineering. There are several reasons for this. One is the fact that situated problem solving areas are site specific. Architectural design in particular aspires to be integrated with its physical context. This requires that the site of a building be represented in a variety of forms, through contour maps, photographs, 3D sketches, traffic count histograms, and computer models. Furthermore, these representations remain relevant throughout the course of the design process. During the early stages of design, the contour map may play a central role especially if the terrain on the site is steep or predominant in some other way. At the end, the completed design may be presented to the occupants of the building through a model, as it would appear on the physical site (as is the case in many notable, public buildings).

Furthermore architecture is socially situated. In this regard, it shares common features with industrial design and information systems. All products are designed to fit a social context to varying degrees. A jet engine is visible to the public only from a distance, yet at times, it comes into close contact with humans, such as, in the case of the maintenance crew. Its design, however, is by and large a response to its technical requirements and not to its human interface. In contrast, a toaster is designed to respond to ergonomic needs of the user. More critically, a computer not only has to respond to ergonomic needs but also to the cognitive needs of the user.

Architecture, like these later forms of design, is intended to accommodate the user along many dimensions: functional, psychological, cognitive, ergonomic, climatic, economical, and so on. Humans live around architecture but also in and on them. The behavior of the users is an integral part of the functionality of the object. Living in a building means that there is a cognitive, ergonomic, psychological and economic interaction with it on a continuous basis. No other artifact of human design can claim a similar range, scope, and directness of use by humans. Consequently an architectural design problem gets to be represented through a variety of man-environment interaction parameters: demographics of the user, their habits of interaction, clothing, privacy needs, and visual perception.

Architects then are faced with handling situated, multi-faceted, and multi-media driven representations that tend to persist throughout the entire design process. Unlike other disciplines the introduction of new and diverse sets of representations are encouraged and is even seen as the bane of the architect's creative endeavor. *Thus architecture is a representation saturated problem domain*, more so than any other with which I am familiar.

2.2.3 ***Alternative Generation***

In all fields of design a distinction between routine and creative (or innovative) designs can be made. Incremental improvements of a prototypical aircraft design to achieve faster and safer rides or marginal variations on the prototype (fewer passengers with a greater flight range) normally require acts of routine design. Conversely, coming up with watershed designs, such as the first aircraft, the first aircraft to break the sound barrier or to escape earth's gravitational pull, require a different kind of design process.

Without trivializing routine design, the distinction I am making here has to do with well-defined design problems versus ones that are ill-defined. Well-defined problems are characterized by the clarity of their initial states, methods of generating and evaluating proposals and goal states. Ill-defined ones (Eastman 70, Simon 73) on he other hand, are without clear goals, points of departure or methods. There is an interesting relationship between *non-routine* and *creative*, as related to this notion of ill-defined, which I will tackle a little later.

All problems, including design problems, have some structure. Their goals, for example, may be stated simply as "designing a house for a client." If this were a routine problem (which it sometimes is) the architect would pull a stock design from her shelf and supply it to the contractor to build. This is what tract housing is like and this is what contractors working without architects tend to do.

In some instances however this approach does not work. The client, presumably, needs something more because she realizes that her site is unique and so are her needs. The architect must begin by further defining her goals. She visits the site. Interviews the client and reaches into her repertoire of professional skills. She decides that this will be a sustainable house, or that this house will stand out from its neighbors, or that the it must be placed on top of the falls rather that facing them (as one very famous house goes). This type of goal clarification invariably shifts the range of design methods that can be used. These may include finding sustainability techniques applicable to this site, the latest formal styles used in architectural circles, or a careful study of the physical amenities of the site. Finally, the test for the success of the design is influenced by these very goals that have been redefined, in addition to larger performance issues like technical feasibility, schedule of completion and conformance to the budget.

Ill-defined problems require this kind of redefinition of goals, means and ends as a matter of routine. Within such a scenario of redefinition, stability can be hoped for but not guaranteed. In other words, initially redefined problem goal, say sustainability, may not remain valid throughout the design process. While exploring a means for achieving this end, the designer may decide that this is an unattainable goal or that real sustainability costs too much. In the end, the means that are available to the designer may help redefine the goal of the task. For example, symbolic sustainability, which illustrates nature and natural things, may be used to elevate users' consciousness, without actually making the building sustainable.

Designers in every field realize these distinctions and behave accordingly. When a routine problem is given they rely on well-understood strategies. When creative designs are needed they find ways of breaking the routine established by these well-known *frames of reference* (Akin 96) and search uncharted territories of design.

Architects, designing in laboratory conditions, have been observed to apply creative design strategies even when the problem is manifestly not one that requires such responses. In a recent experiment we conducted, we asked architects to arrange a given set of office furniture in three alternative, fixed spaces (Akin 86a). The furniture and the spaces were represented as in-scale, cardboard cutouts. There was clear control of the initial problem representation and what could be manipulated during design exploration. The space provided for the layouts was so restricted that in two cases, L-shaped and square-shaped layouts, the feasible solutions were less than a hand-full. In the third case, the rectangular layout, even one solution was barely possible.

On the average, during the course of 30-minute design sessions experienced *architects* generated 4.13 solutions against the 1.83 generated by *students* of architecture and 1.253 by *novices* (Table 1). These measurements also correlated with the difficulty of problem.

Novices, students and architects generated more solutions on average for the rectangular layout than for any of the other two layouts.

Table 1: Average number of re-structuring of the problem (Akin, 1986a).

Subjects (n = 18)	**Rectangular Site**	**Square Site**	**L-shaped Site**
Architects (n = 6)	4.00	4.00	4.33
Students (n = 6)	3.50	1.00	1.00
Novices (n = 6)	2.50	1.00	3.00

What is interesting in these findings is that architects generated new solutions even when they had already found satisfactory ones. This was not the case for novices or students. The performance of the latter was markedly in favor of fewer *restructurings* of the problem (redefining the constraints of the problem space). Once a solution was found that was it. Particularly under the more difficult layout conditions (for example the rectangular shaped site) their restructurings occurred only in search of a single, feasible solution.

The questions are "why do expert architects restructure problems and launch new searches even beyond the point of finding a satisfactory solution?" And "why is this behavior absent in students and novices?" I believe that this has to do with the training and design habits of architects. Architectural education is preoccupied with creative solutions. Students are urged to find them. Those who do are rewarded. Those who do not are penalized. Routine solutions are equated with the mundane and creative ones are equated with the sublime. Heroes and exemplary designs always fall under the category of the hi-style, master architect and their products. Architects' heroes are the super-performers of their genre, while ordinary performers are relegated to the rank of the anonymous trooper. In other words, expert architects are conditioned to use *creative-design* strategies under all conditions, even when the problem does not warrant it.

Ball, (97) studied the engineering design process by observing electrical engineers attempting a novel integrated-circuit design problem. In the protocol data they collected, they found "i) a failure to search for alternative solutions, ii) a marked inclination to stick with early 'satisficing' solution ideas even when these were showing deficiencies; and (iii) only superficial modelling and assessment of completing alternatives when such options were actually considered."

Table 2 shows the number of initial concepts, moderately or substantially different alternatives and minimally different alternatives generated by six subjects for hardware solutions (numerator) and mathematical solutions (denominator). The number of alternatives generated in either category is remarkably low. Based on these observations we can state that these engineers tend to apply routine-design strategies even when the problem calls for a novel design solution. They stick with only a few (often just their initial) problem definition. They push the solution through known methods hoping to satisfy the problem almost by brute force.

Table 2: Number of initial and alternative solution concepts generated at the hardware/mathematical design levels (Ball, *et.al.*, 1997)

Category of Solution Concept	DS	JO	IH	JF	JM	JC
Initial Concept	32/5	45/8	23/11	26/5	23/10	7/7
Alternative Concept (different)	2/1	1/0	0/0	1/0	2/1	0/0
Alternative Concept (min. diff)	2/3	1/3	1/2	0/0	0/1	0/2

Altman *et.al.*, (99) in their study of engineering students doing engineering design found that the "number of alternative solutions considered" did not correlate with years of engineering education. However, they found positive correlation between the number of alternatives and the quality of the design solution.

Are these behaviors endemic to the domains in which they have been observed? Are they typical or are they idiosyncratic? It is difficult to say with this amount of data we have but the setting of a relevant hypothesis is plausible: architects tend towards creative design strategies while engineers tend to routine design.

2.2.4 Problem Composition

In almost all design disciplines there are standard ways of decomposing a design problem. They break down the design object into mutually exclusive constituent sets. The elements in these sets are designed individually and then they are combined into larger and larger solutions. There are two aspects of this process that create invariants for design fields: (1) the schemata used for decomposing, and (2) strategies used for recomposing. The former will be dealt with in this section. The latter will be discussed in the next section.

3 Design Phases

Frankenberger (97) describes one of the most comprehensive forms of a decomposition schema used in mechanical engineering and in different forms in industrial design. The problem is broken into discrete design sub-problems in a three dimensional space where the dimensions are: design steps, sub-components of the design and solution variants. All of these factors then play out along the fourth, time dimension, shown in Figure 2 as a snaking path in the three-dimensional space.

In studying the behavior of architects researchers have shown that architects too decompose their problems into sub-problems, develop variants and follow design steps. Yet none of

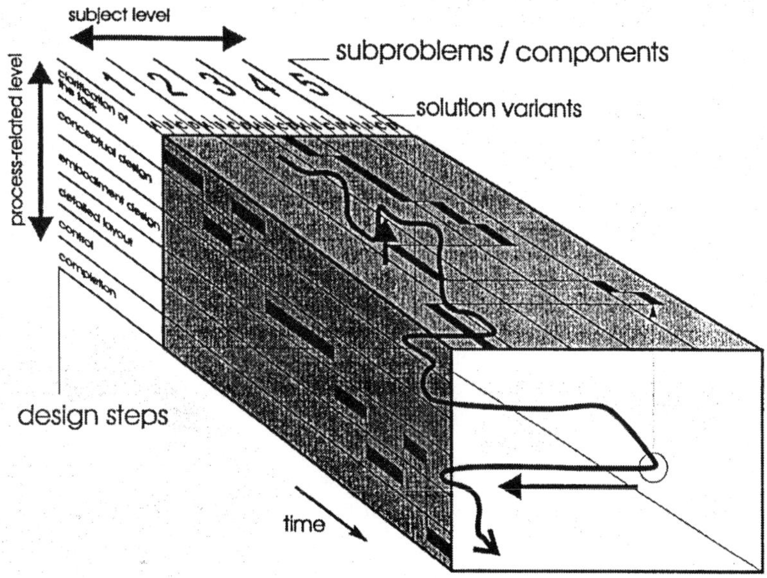

Figure 2: Description of a design process by sub-problems, variants, and design steps (Frankenberger, 1997)

these schema have been standardized in a form that satisfies the requirements of most architects. The best that we can do, as architects, when asked about the stages of design is to refer to the standard project phases used in professional practice. These usually follow the recommendations of professional organizations like the AIA and RIBA, which are generally useful for scheduling deliverables and payment schedules. There is little in the way of creating a useful normative stage model for architecture. It is up to the individual architect to define information processing stages for her design process.

Design protocols show that industrial designers use schemata such as: programming, data collection, analysis, synthesis, development, communication (Dörner 95). Mechanical engineers have guidelines that include even more specific steps. For example the VDI Guideline 2221 stage model suggests a sequence that consists of: clarify and define task, determine functions and their structures, search for solution principles and their combinations, divide into realizable modules, develop layouts of key modules, complete overall layout, prepare production and operating instructions, product documents, and further realization (Eisentraut 97).

4 Constituent Design Problems

Similarly, in architecture, the decomposition of larger problems into smaller ones is not a straightforward proposition. In other fields, especially in the sciences, knowledge domains are defined through standard problems that identify the sub-parts of the domain: such as integration and differentiation in calculus, kinetics and statics in physics, organic and inorganic materials in chemistry and so on. Each of these categories can be further decomposed into finer grain problems, until a desired disaggregation of topics is obtained. This is not a straightforward or trivial task. In fact, the discovery of the periodic table by Mendelev is attributed to performing just such a task: creating a hierarchical constituent list for a chemistry text he was in the process of writing (Kedrow 66-67).

Akin and Hayes (99) have attempted to find a similar structure in the knowledge domain of architecture. They start with the assumption that there are two potential beginning points for structuring this domain: building types (hospitals, museums, schools) and building component systems (enclosure, structure, circulation). Upon further examination of the field it became clear that the latter was dependent on the former more so than the other way around. In other words, given a building type, it was easier to determine which component systems were more critical and how choices could be made between these systems. Consequently they selected a building type (museums) and asked architects and educators to create classification schema for it. Subjects were given a comprehensive list of 120 museums, documented in several books. These were chosen to exhaust, as completely as possible all of the significant examples that have been constructed in the 20th century, including additions to older ones such as the Louvre and the National Gallery of Art, in Paris and Washington, DC, respectively.

To aid the classification task subjects were given cards identifying each building with an iconic picture and its name. They were asked to place the cards into piles and sub-piles based on their criteria of classification. Then they were asked to write down the criteria that they used to classify each building. It was expected that this would yield the constituent hierarchy of the topics and sub-topics that are relevant to all of the buildings under this building type. The categories used by different subjects included titles such as: transformed building, many parts juxtaposed, addition to existing, rooms around center, grid/orthogonal, linear/route, ornamented, "look at me", sculptural, linked to urban context. It is clear that neither a shared taxonomy nor a consistent lexicography would emerge from this data. For example, while *concentric* and *linear* refer to topological characteristics, *urban context* refers to the site of the building and *transformed* refers to the process of creation of the design.

A preliminary analysis of the first handful of subjects, however, revealed an interesting hypothesis. When the content of the categories are compared between subjects, there is some agreement. On the average, there are 28 overlapping categories developed with an average of 78 items shared between these categories, across subjects. There are at least eight groups (out of the total of 28) that shared four or more entries between any pair of subjects. Further analysis is necessary to find groups globally recognized by all subjects.

In conclusion, architects seem to agree on a small but non-trivial number of problem categories that apply to a given building type but, on the average, there is at least as many errant classifications between subjects. Furthermore, there is very little agreement between subjects about the lexicographic descriptions of these categories.

5 Breadth and Depth-first Search

What, then, can we say about the architects search domain as structured by the phases and constituents of the design problem? It is clear that architects are able to sensibly categorize a large number of buildings into constituent hierarchies. This, I postulate, provides the sub-problem schemata for structuring the architects search domain.

In an earlier experiment Akin (86b) showed that designers structure their search domains by major locational options (i.e., site location) against different organizational categories (such as, plan parti, topology, orientation, geometry). First, architects search *breadth-first* along the principal options. Then they develop one of the options in detail, *depth-first*, until all design categories are exhausted. I call this process *depth-and-breadth-first* search, a hybrid of the classical search strategies used in problem solving.

5.5 Complexity Management

As we saw in the findings of Ball (97), electrical engineers do not look at major alternatives before proceeding with one of the selected options. Further, architects do not follow a process driven by a procedural schema (such as, analysis, synthesis, evaluation) as represented in the normative design models of industrial and mechanical design; rather their process is driven by the product and its breakdown into a constituent hierarchy.

Architects work iteratively. In the realm of design automation, this is often referred to as a weakness. Whereas in cognitive terms it is a human way of making complexity more manageable. The decomposition of the design problem, which has been described above, is only one facet of this. The other even more crucial aspect of dealing with complexity is the recomposition of a comprehensive design solution from partial ones.

Given that architects decompose their problems into coherent constituent hierarchies (at least for themselves), how, then, do they re-assemble these partial solutions into singular solutions. In earlier research, architects were asked to design single-occupant residences (Akin 94). Figure 3 illustrates the space of partial solutions that are developed during the

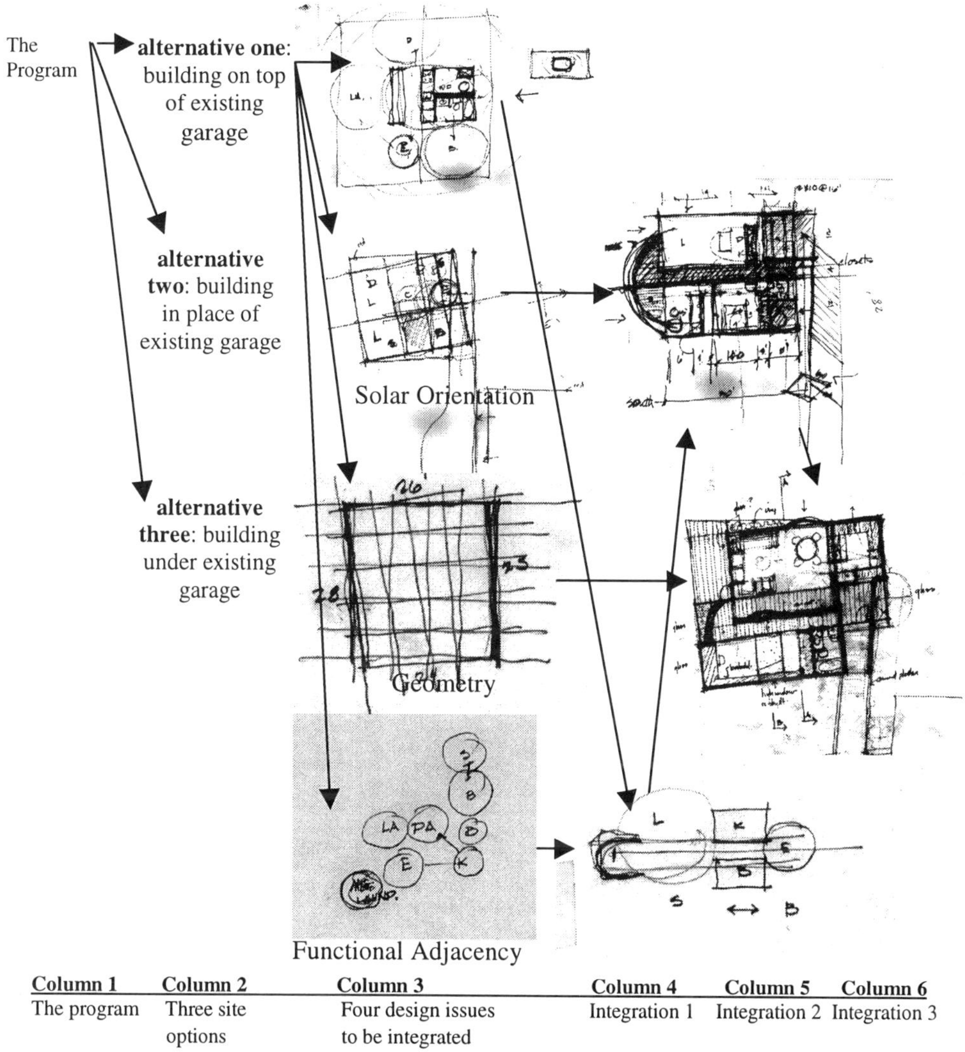

Figure 3: Pairwise integration of partial architectural design solutions (Akin, 1994)

course of one of these sessions. The node on the far left of the graph represents the problem brief and site information provided to the architect. The next column of nodes represents the three alternative locations considered by the architect: above-grade solution, on-grade solution, and subterranean solution.

Only one of these, the above-grade solution, is further developed in this design session. The remainder of the graph represents the categories for which partial solutions are developed. These consist of the *design concept, solar orientation, geometry,* and *functional topology.* In column 3 of Figure 3, the partial solution for the design concept is shown as a simple diagram representing a unit with a wet-core. The node, solar orientation, located below it in the same column shows that this partial solution is achieved by placing the daytime functions of the unit along the West and South sides and the night time functions along the East and South sides of the site. The irregular geometry of the site is resolved by overlapping two 24'x24' squares of 6'x6' grid patterns each, shown in the next node labeled geometry, below this one. This is seen as the planning grid incorporating the irregular site geometry into a more regular shape. The last node of column 3 labeled functional topology shows a bubble diagram, which places the various elements of the program in their proper relationships to one another. Each of the nodes of this column represents how individual criteria are satisfied through partial solutions. None, alone, however, represent the comprehensive design solution.

The comprehensive solution to this problem emerges at the very end of this 4-hour design session and is shown in the subsequent nodes to the right of column 3 in Figure 3. The links between these nodes show how this is accomplished. In the fourth column of nodes the concept and topology solutions are merged. Then this solution is merged with the solar orientation solution. In the next column, the architect begins to test the idea of merging geometry and dimensions into a yet new solution. This test fails. The design turns out to be too large for the site. A new solution is developed by matching the current layout with the outline of the whole site. The sketch in the final column shows how the architect manages to integrate all partial solutions into a single comprehensive solution.

The strategies used to integrate these partial solutions are not standard. In other words, there is no known procedure as to how a topological solution is integrated with geometric parameters of the site, or how a plan parti can be merged with a solar orientation based layout. Unlike the parts of a VLSI circuit or a mechanical assembly, the interaction of the parts in an architectural design are not theoretically determined. The architect devises *ad hoc* strategies for accomplishing pairwise integration. This is clearly another aspect of architectural design that differs form the others.

6 Conclusions

During the course of the discussion presented above several questions pertinent to the variants of architectural cognition have been raised. Below I will attempt to respond to these questions that are based, more or less, on the evidence provided above.

6.6 What do architects do that others do not under similar conditions?

We have considered several important differences between architects and other designers:

Architects rely on *analog* representations as a central feature of design discourse. Even though they also use *symbolic* representations, the integration of partial solutions into a singular solution is achieved in the context of analog representations.

Architects use a greater variety of representations and for a longer duration during the design process. Due to the situated and user dependent features of architectural design problems a plethora of representational formats and media are required. Lack of universally accepted representational standards contribute to the complexity of this picture.

Architects, while exploring solutions, rely on a greater number of alternatives than other design professionals. It is possible that this is influenced, if not caused, by the emphasis and value architectural education places on creativity and unique designs. Architects continue to search for alternative solutions even when they have already developed satisfactory ones.

Phases of the design process in architecture do not follow a prescribed staging. Architectural problems are decomposed into sub-problems idiosyncratically rather than based on a globally adopted schema of how a building's constituent parts ought to be structured. As a result, architects seem to search first in breadth, developing major alternatives in order to structure the problem domain, before they search in depth with one of these principal alternatives.

Architects use a pairwise integration strategy to recompose the partial solutions generated into a holistic design. For example, the topological arrangement of functions on each floor can be matched with compositions of solids and voids on the building facade, or the facade may be integrated with a geometric constraint imposed by the site. These strategies are driven by designers' personal knowledge and skills, which have been developed through their own experiences.

6.7 What cognitive activities remain constant across tasks and individuals?

The architect deals with a greater variety of tasks using a greater variety of methods than do other designers working in other areas. There are local standards, within certain practices and for certain building types, but architects rarely ever specialize to the exclusion of a diverse set of building types, primarily for business reasons. Even when they do, site and climate conditions demand diversity in the solutions that they have to consider. This applies to high style architects as well as small offices that dominate professional practice.

As a result, I conjecture that individual designers are influenced by to the treacherous terrain of unmanageable, ill-defined and unruly problems in which they have to cleverly maneuver their skill banks to achieve satisfactory or creative solutions. Understandably, they rely on strategies that resemble those of *bounded rationality* (Simon 47) in which scarce cognitive resources are stretched to cover these unmanageable conditions and sometimes with undesirable results. A small example of this is found in the fluctuation of the number of alternatives generated by subjects in response to the difficulty of problems (Akin 86a).

In larger offices where the overall portfolio may be diverse, individuals become specialized by building type or design phase (conceptual design, programming, quality control, etc.). This is where some of our observations about the design process may in fact not apply.

6.8 What justification is there to have a common core for different design disciplines?

I believe the difficulties in creating a common core spanning multiple design disciplines are serious ones. Designers are notorious for preferring situated learning. They go from instance to case using an informally developed bundle of design principles (March 76). The first barrier that must be overcome then is to find a shared set of problems that appeal to multiple disciplines that are to be integrated in a core course set.

The second problem is to bridge the gap between the idiosyncratic nature of architectural knowledge and the regularity of problems in the domain. Standards of problem decomposition and design staging is necessary, if a common problem set is to be used to drive the core course set for education. Educational programs and strategies may in fact spur on the emergence of such a problem set. In other words the core course set may be the prophecy that can fulfill itself. This however requires a level of maturity in the objectives architects seek in their designs. So long as there are self-declared prophets of architectural style that indulge in esotericism and flocks of students follow them closely, such a consensus is unlikely to emerge.

References

Akin Ö, (1986a), A formalism for problem structuring and resolution, *Environment and planning b: Planning and design,* 13, 223-232.

Akin Ö, (1986b), *Psychology of architectural design*, Pion Ltd, London.

Akin Ö, (1994), Psychology of the early design process, proceedings of the *design decision support systems conference*, Vaals, The Netherlands, August 15-19.

Akin Ö, and C. Akin C, (1996), Frames of reference in architectural design: Analysing the hyper-acclamation (A-h-a!), *Design Studies,* 17, 341-361.

Akin Ö, and Akin C, (1997), On the process of creativity in puzzles, inventions and designs, *Automation in Construction*, 7, 123-138.

Akin Ö, (1999), Course notes: *48-510 Decision making in architecture,* School of Architecture, Carnegie Mellon University, Pittsburgh, PA 15213, Spring.

Akin Ö, and Hayes J R, (1999), Unpublished study on "Architects' problem classification." Center for Innovation in Learning, Carnegie Mellon University, Pittsburgh, PA 15213.

Atman C J, and Turns J, (1999), Studying engineering design learning: A case study in the application of protocol analysis and the development of measures, Proceedings of the conference on *Knowing and Learning to Design,* supported by Edutech Institute, College of Architecture, and College of Computing, Georgia Institute of Technology, Atlanta, GA, April 27-28, 1999.

Ball L, Maskill L, Ormerod T, (1997), Satisficing in engineering design: Causes, consequences and implications for design support, *Automation in Construction,* 236.

Chase W G, and Simon H A, (1973), The mind's eye in chess, W.G. Chase Ed. *Visual information processing,* Academic Press, New York, NY, 215-282.

de Kleer, (1975), *Qualitative and quantitative knowledge in classical mechanics,* Master's Thesis, MIT (December).

de Kleer, (1977), Multiple representation of knowledge in mechanics problem solver, *International joint conference on artificial intelligence*, 5, 299-304.

Dörner D, Ehrlenspiel K, Eisentraut R, and Günther J, (1995), Empirical investigation of representations in conceptual and embodiment design, V Hubka Ed. *Proceedings of ICED 95*, Edition Heurista, Zürich, 631-637.

Eastman C, (1969), Cognitive processes and ill-defined problems: A case study from design, Proceedings of the international joint conference on artificial intelligence, Washington, DC.

Eisentraut R, and Günther J, (1997), Individual styles of problem solving and their relation to representations in the design process, *Design Studies,* 18, 369-384.

Frankenberger E, and Badke-Schaub P, (1997), Modelling design process in industry-empirical investigations of design work in practice, *Automation in Construction,* 243, 1-17.

Goel V, (1992), Ill-structured representations for ill-structured problems, Proceedings of the fourteenth annual conference of the cognitive science society, Lawrence Erlbaum and Associates, Hillsdale, New Jersey, 844-849.

Goldschmidt G, (1991), The dialectics of sketching, *Creativity Research Journal,* 4:2, 122-143.

Larkin J, and Simon H A, (1987), Why a diagram is (sometimes) worth 10,000 words, *Cognitive Science*, 1165-99.

Kaplan C, and Simon H A, (1990), In search of insight, *Cognitive Psychology,* 22, 374-419.

Popper K, (1963), *Conjectures and refutations: The growth of scientific knowledge,* Routledge & Kegan Paul, London.

Kedrov B M, (1966-67), On the question of the psychology of scientific creativity, *Soviet Psychology* 5, 2, 16-37.

Larkin J, (1983), The role of problem representation in physics, *Mental models,* Ed. by D. Gentner and A. L. Stevens, Lawrence Erlbaum and Associates, Hillsdale, New Jersey.

Levin P H, (1984), Decision making in urban design, N. Cross, Wiley, Chichester Eds. *Developments in design methodology.*

March L, (1976), Introduction, March Ed. *The Architecture of form* L. Cambridge University Press, Cambridge, England.

Newell A, and Simon H A, (1972), *Human information processing,* Prentice Hall, Englewood Cliffs, New Jersey.

Simon H A, (1973), The structure of ill-structured problems, *Artificial intelligence,* 4, 181-201.

Chapter 7
Defining Design between Domains: An Argument for Design Research á la Carte

Craig Zimring and David Latch Craig
College of Architecture
Georgia Institute of Technology
Atlanta, GA

Abstract

A domain-independent theory of design could provide a common framework for design education and research. However, concepts such as "domain" and "design" have yet to be clearly defined. This chapter explores common descriptions of design – that design involves abductive reasoning, construction, ill-defined problem solving skills, or wicked problem solving skills – and argues that they are not always sharp enough to both distinguish design from other types of problem solving and unite design across different design-related disciplines. It is proposed that research and education focusing on mid-level constructs, such as analogy, may help to solidify design research by yielding results that cross design disciplines.

1 Introduction

The continued growth of multidisciplinary design conferences, books and journals raises the very attractive possibility of creating a domain-independent theory of design, that is, one that is useful to a broad array of design disciplines including engineering, architecture, computer science, industrial design, planning and even the performing arts. Some authors have proposed this directly (e.g. Cross 95; Cross, this volume), while others have assumed a domain-independent theory, in some cases taking data from different design disciplines to draw conclusions about design in general (e.g. Gero and McNeill 98; Goel 95). Such a theory could provide a common framework for researchers in different disciplines, making it easier to share findings and to improve teaching methods and design aids. This is of pragmatic importance, partly because of the tenuous nature of design research in different academic areas. As it stands, traditional areas of research outside of design, such as the natural sciences and some areas of engineering, do not always consider design as distinct entity worthy of independent research. They may treat design as non-academic or at best as the realm in which their research is applied. A domain-independent theory of design, with its own design-specific research practices, might help to legitimize studies of design within an academic setting and thus draw more researchers into its fold. The question, however, is whether current discussions suggest a useful direction for a domain-independent theory.

In this chapter we briefly review the ways the discussion has been framed and the ways key terms such as "domain" and "design" have been used. The chapter continues with an examination of common claims made about design, such as that design involves abductive reasoning or that it involves solving ill-defined or wicked problems. Of special interest is whether existing descriptions set clear and useful boundaries, or whether they muddle similarities between different design disciplines and differences between design and other types of activity. Ideally, a domain-independent theory of design will accommodate differences between domains without limiting the theory's application in any overt way. Finally, we argue that while existing perspectives on design are illuminating, none seem sufficient for articulating a domain-independent theory of design. We suggest that studying mid-level process in design, like analogy, may be more profitable in scaling research across disciplines.

2 Defining domain

Domain independence in design usually means at least two things: 1) that there are meaningful commonalities across instances of design in different disciplines (Cross 95), and 2) that design is distinct from other kinds of problem solving. In this section, "domain" is considered in more detail. First it is considered as a general research construct in cognitive psychology and second, as a possible way to organize design research across different design disciplines.

2.1 Domain as a research construct

The concept of *domain* escapes easy definition. Some researchers may use a priori categories that are of general interest, such as different disciplines. In discussing domain specificity in cognition, Frensch and Buchner (99) argue that the key to the debate is the generality of cognitive constraints that affect observable cognitive behavior. Specifically, they argue that a domain is "anything that a given constraint can potentially be generalized to and from" (p. 142). Hence, the observation that expert knowledge is arranged hierarchically suggests that expertise is a domain (Chi 89). Similarly, domains may be pictures versus words, children versus adults or letters versus numbers (Frensch and Buchner 99). In this case, what defines domains are features of tasks or of agents carrying out tasks.

The purpose in examining domain boundaries is to help illuminate the underlying processes and decide between competing explanations of behavior. Differences between memory for numbers and memory for letters, for example, does not mean that research on one needs to be insulated from research on the other; rather it suggests that studying the difference between numbers and letters themselves (how they are learned, how they are used, their logical structure, etc.) may yield insights into memory in general. Although this approach to using "domain" as a research construct provides considerable flexibility given the variety of features that may be associated with human behavior, it also appears somewhat circular: The exploration of domains defined by given features is meant to help define the domain itself. If, for example, we compare different types of *design* (presumably by looking at different types of tasks and different types of problem solvers), we seem to be assuming that there is in fact some superordinate category called *design* that can be informed by such research. Problematically, circularity invites the miring of normative and descriptive accounts of design behavior. One alternative, which will be returned to at the end of the chapter, is that design in different disciplines represents different types of problem solving, not different types of design, and that design in the large is at best an illusory category.

2.2 Domain in design

Some explorations of design have used the discipline of practice – for example, architecture vs. engineering – as an independent variable (e.g. Purcell and Gero 96; Roozenberg and Cross 91). In this respect, design disciplines have been explored as domains of behavior much like numbers and letters have been explored as domains in memory research. In this volume, Akin, for example, examines differences between engineers and architects, concluding that there are observable differences in problem decomposition and other aspects of their behavior. Although insightful, such studies focus primarily on disciplinary labels, hiding more complex differences that exist between the participants in their studies. Although disciplinary labels represent natural, easily grasped distinctions, differences in observed behavior are clearly not inherent in the labels themselves. Rather, they stem from differences in knowledge, procedural skills, cognitive processes, general problem structures and social practices. The complexity of these differences ultimately limits the extent to which research across design disciplines can help inform our understanding of design in general.

Rather than marking domains simply by using discipline as an independent variable, one might construct a featural description of design where the features that differentiate related design activities are teased apart and opened for inspection. There are, for example, four areas where potentially large differences in what designers do may crop up: differences in practices that are responsible for giving meaning to and shaping design activities, differences in the sorts of tasks designers face, differences in the kinds of artifacts produced by designers, and, finally, differences in the kinds of knowledge used and in the cognitive processes carried out.

Differences in practices are influenced by educational, professional and corporate cultures as well as by larger social, economic and regulatory forces. Aspects of practice include, among other things, the context of the design activity, strategies, physical activities, social activities, standards of rigor, and the types of external representations favored by each discipline. External representations used differentially across different types of design include language, math, orthographic projections, orthographic sections, renderings, three-dimensional models, charts, flow diagrams, and topological graphs (see, for example, Akin, this volume; Eastman, this volume). Other differences in practices include differences in attitudes towards the use of knowledge in design. Roozenburg and Cross (91) argue, for example, that engineering disciplines that have a history of being explicitly knowledge intensive (i.e., analytical) may now see themselves as being in a position to fold more design skills into their knowledge, and particularly to admit the value of intuition in suggesting solutions early in the design process. Disciplines such as architecture, on the other hand, come from the opposite perspective where the focus is more on incorporating analysis and structured procedures into intuitive practices.

Differences in tasks, on the other hand, include the fact that design problems in some disciplines demand more accountability and precision than in others. An architect who designs a vertical circulation system in a building may, for example, have less at stake than the engineer who designs its structure. Another major difference between disciplines at the task level lies with task constraints. Some disciplines – engineering, for example – may favor the use of standard parts and production processes because of the costs associated with production setup. Architects and software designers, on the other hand, face such constraints to a much lesser degree. They may conform their designs to standard building or programming practices respectively, but major design components may be uniquely structured simply because production setup costs are low or because existing production processes are flexible.

Differences in the artifacts specified by designers include differences in form, modality, permanence, scale, use and other qualities. Many aspects of buildings such as layout and appearance are at least theoretically understandable by end-users during design, resulting in patterns of collaboration, testing and accountability that differ significantly from those associated with more "invisible" design processes. An engineer, for example, working on a car engine is not likely to collaborate with end users directly, given the difficulty the average car driver has in understanding the mechanics of engines.

Finally, differences in knowledge may include differences in the organization of information, differences in the modality of encoded experiences and differences in the degree to which knowledge is encoded in abstractions, analogies and exemplars. Differences in knowledge may ultimately derive from differences in tasks and practices. If a task focuses on human experiences rather than on, say, mechanical performance, there may be a tendency to retrieve and abstract previously unanalyzed personal experiences (Downing 92) or gestalts (Schon 88) rather than explicit procedures, schemas or rules (Chi et al. 81), regardless of expertise (Rambow and Bromme 95).

Clearly, this only scratches the surface of the differences between different types of activities generally called design. Given the variety of features associated with design activities, disciplinary labels are limited in the degree to which they represent actual differences. Moreover, while there are significant differences between disciplines, there are also significant differences within them. A software designer, for example, could be working on a piece of code that controls a cooling tower for a nuclear reactor or, alternatively, on a game for kids, each of which may fall on opposite ends of the task, knowledge and practices spectrums. Given the many ways design activities can differ, along with the fact that the differences themselves are hard to measure, exploring the degree to which different features of design-like activities mark cognitive boundaries may be futile. The next section considers a different approach, one that involves defining design outright rather than exploring domain boundaries empirically.

3 Defining design

In seeking a definition of design, common English dictionaries are not much help. For example, the current Webster's Dictionary lists the verb design as "to create, fashion, execute, or construct according to plan" and the noun as "the creative art of executing aesthetic or functional designs" (Britannica 00). For some theorists and methodologists, the study of design begins with the observation that design problem solving is distinct from other human activities but nonetheless prevalent in everyday life. Schon (92), for example, proposes that all teachers and students are designers because they make "things out of the materials of a situation under conditions of complexity and uncertainty" (p. 23).

Simon (81) famously defined design as the act of "transforming a given state of affairs into a desired state of affairs." He also argued that there is something universal, and apparently fundamental, that can be exchanged between designers working in different areas, whether that thing is analyzed or not. A music composer and an engineer, on this view, "can carry on a conversation about design, can begin to perceive the common creative activity in which they are both engaged, can begin to share their experiences of the creative, professional design process" (p. 158). The real question, however, is whether a music composer and engineer could actually carry on such a mutually rewarding conversation. When design is abstracted into something like, 'transforming a given state of affairs into a desired state of affairs' there is no doubt that designers from different disciplines and backgrounds share something in common. Others have produced similar definitions for the sake of unity. Mistree et al. (90) propose, for example, that describing design strictly in terms of decisions would yield "a description of the process written in a common 'language' for teams from various disciplines – a language that can be used in the process of designing." Such statements about design, however, are relatively simply and, as such, are not likely to support much of a conversation at all. Invariably, a conversation would uncover more features, attributes and patterns of design, all of which would increase the chance that disagreements would eventually arise about the subject being discussed.

How, then, should one go about framing design problem solving in a domain-independent fashion? As one approach, design could be framed as a distinct set or sequence of psychological processes. Some researchers, for example, have focused on identifying the basic cognitive processes and constraints that characterize design. Design, in this case, might be understood by merging task-level constraints (i.e. "boundary conditions" of cognition) with internal constraints (e.g. short-term memory and serial processing). Using such an approach, Newell and Simon (72), for example, developed the General Problem Solver to enact a domain independent theory of problem solving, requiring only the translation of domain-specific representations into computer-specific representations. General problem solvers, however, have never been very successful with anything much tougher than toy problems, which in turn may suggest that a domain-independent model of design or even problem solving in general is untenable (Frensch and Buchner 99).

On the other hand, fixing a general model of design may simply require that we better understand the organization of knowledge and the strategies used in different disciplines rather than overhauling or abandoning existing models themselves. Cross (95), for example, hints that there may be ways of organizing and operating on knowledge that are unique to design – that, in other words, there are "particular, 'designerly' ways of knowing, thinking and acting." Goel (95), in a similar vein, points out numerous features of design, both at the task level and the psychological level, that Newell and Simon failed to incorporate into their model. Along these lines, the ideal of a general model might be achieved by focusing more closely on what designers do.

Defining design by specifying what makes design behavior unique effectively inverts the strategy discussed in the previous section. Rather than exploring the boundaries of design, design is considered from the inside-out, that is, from its most essential features to those which may be shared with non-design activities. Such efforts usually begin with what makes design different from other activities and other types of problem solving, while striving to incorporate features that are shared across different design disciplines. Looking at how different types of activities generally called design are similar may, of course, result in the same growth of features that plagues attempts to explore domain boundaries. However, usually lurking beneath featural descriptions are explanatory theories of the things being described (Murphy and Medin 85). In this section we examine four fairly coherent explanatory theories popular in framing design research: design as abductive reasoning, design as ill-defined problem solving, design as wicked problem solving and design as construction. The purpose of the section is to explore the extent to which these explanatory theories help in establishing a domain-independent theory of design using the two requirements listed earlier, that they unify different activities commonly called design while distinguishing design from other activities.

3.1 Design as abductive reasoning

Design theorists have long attempted to characterize the logical reasoning processes in design (e.g. Asimow 62). Part of this effort has addressed the extent to which design is similar to science and, in particular, the extent to which design can be seen as relying on two modes of logical reasoning: induction and deduction. Using induction, on the one hand, a design engineer working on a building might infer which structural features of buildings improve performance in high winds based on a history of structural performance. Using deduction, on the other hand, the same design engineer might reason that a specific building with certain structural features will, in fact, perform well in high winds. Problematically, design is not easily captured by any simple combination of these two processes. In particular, design appears to turn deductive reasoning on its head. Rather than inferring performance characteristics from specific instances, designers use performance characteristics to *generate* specific instances.

The difficulty in design is that when deductive procedures are reversed, they are not deterministic. When designing a building that will be stable in high winds, for example, one is free to imagine structural features that have never been imagined before and hence would not be part of any prior inferences. To illustrate, when the architect Hugh Stubbins and the structural engineer William LeMessurier designed the structural system of the 59-story CitiCorp center in New York they used an innovative chevron-shaped wind bracing. Among other things, this allowed them to construct a free-standing church under the building. Mismatch between the reasoning that went into their innovative structure and the logical analysis of buildings at that time, however, became apparent when an architecture student pointed out that the building was susceptible to failure in seventy-mile-per-hour quartering winds. Initially, the structural system was evaluated using traditional assumptions about wind forces, which are based on considering the forces as the total of the vectors directly perpendicular to the faces of the building. Although suitable for structures predating the CitiCorp center, this assumption broke down on LeMessurier's unique design, ultimately demanding a multi-million dollar renovation (Morgenstern 95).

To the extent that design reasoning does not follow inductive or deductive procedures in any simple way, it has been termed abductive (March 76). Initially proposed by Charles Peirce at the end of the 19th Century, abduction has been used as a way to understand hypothesis generation in science. Pierce argued that explanation sometimes occurs by proposing reasonable hypotheses that can then be tested deductively. Coyne (88) and Roozenberg (92) have argued that design entails a similar process, while other researchers studying creative design have countered logical biases in design research by pointing to the important role that proposed solutions play early in the design process, where full or partial designs go on to influence subsequent refinement of the problem in lieu of purely deductive or inductive processes (e.g. Hillier et al. 72).

At a coarse level of detail, design does not in fact seem fit the mold of induction or deduction. However, it is not clear that abduction presents a particularly useful alternative. Induction and deduction are relatively simple processes and, as such, are fairly well understood. Design, on the other hand, involves many processes; one does not simply reason abductively. At a finer level of detail, it may turn out that induction and deduction do, in fact, play a significant role in design.

Looking more closely at design, there are two possibilities for incorporating deductive and inductive logic into descriptions of design problem solving. One approach follows from the work of design methodologists who describe design as moving from the general and abstract to the specific and concrete, or from analysis to synthesis (Hubka and Eder 96). In such cases, although logic may be inadequate for explaining the synthesis of form, it might be used to organize the context of a design problem (i.e., requirements and constraints at all levels). To simplify the context of a problem, requirements, for example, might be logically organized into a "field description" (Alexander 64) from which the designed artifact can then be derived, in turn minimizing interaction between subsystems. This is consistent with an analysis-synthesis approach that says that if a problem is sufficiently well represented and neatly decomposed, the solution should be obvious (Simon 81). Finally, following the progression from problem to solution, logical reasoning can also be used to evaluate resulting designs.

A much different approach to describing design in terms of induction and deduction involves formalizing design in logic at a fundamental level, or, more specifically, modeling the design process using a symbolic, information-processing system (Simon 81). This view minimally assumes that designers are rational agents. Such a description would not hold, in other words, if designers "were always sloppy fellows who reasoned loosely, vaguely, and intuitively" (p. 134). Although some might feel that designers do reason vaguely and do rely on intuition, the rational problem solving model put forth by Simon goes at least partway in that it accommodates guessing, analogical reasoning and other processes that could lead to dynamic changes in a problem space (Simon 73). One should not then take a logic-based theory of cognition as one that in principle excludes contextual effects, conceptual change or even subjectivity (Vera and Simon 93). Such a theory assumes, simply, that all behavior results from elemental serial operations on discrete, composable and amodal symbols, and that this inner core of cognition happens to be logical.

The fact that Simon (81) spends so much time discussing engineering design and the decision-making processes that accompany it makes it easy to overlook his comments on the erratic and unpredictable aspects of design. Simon does not, for example, limit the function of goals in design to the rational evaluation of alternatives. Goals, he argues, "motivate activity which in turn will generate new goals." In the case of the planned development of downtown Pittsburgh, he points out, for example, that "each step of implementation created a new situation; and the new situation provided a starting point for fresh design activity" (81, p. 187). Simon also seems aware that interaction with the physical situation created by the designer leads to dynamic changes in the task. In the case of oil painting, for example,

> ". . . every new spot of pigment laid on the canvas creates some kind of pattern that provides a continuing source of new ideas to the painter. The painting process is a process of cyclical interaction between painter and canvas in which current goals lead to new applications of paint, while the gradually changing pattern suggests new goals" (81, p. 187).

Similarly, a logical, symbol-crunching processor may also exhibit unpredictable, sometimes even erratic, behavior simply because of dynamic feedback within a complex environment.

Anxiousness to differentiate design from deductive or inductive reasoning may stem in part from the belief that design is distinct from science, both structurally (March 76) and empirically (Lawson 79). March (76), for example, argues that, "To base design theory on inappropriate paradigms of logic and science is to make a bad mistake. Logic has interests in abstract forms. Science investigates extant forms. Design initiates novel forms" (p. 15). This belief may be complicated by the assumption that the rational, logical model of design proposed by Simon is an effort to cast design as a science. By formalizing design along with other types of problem solving, however, Simon is not trying to make design a science but is rather trying to develop a science *of* design, in other words, "a body of intellectually tough, analytic, partly formalizable, partly empirical, teachable doctrine *about* the design process" (81, p. 132, italics added). Formalizing design does not mean neatening it up or making it more efficient, and, more importantly, a formal model that can be interpreted by a computer will not necessarily entail a list of instructions that will be coherent to a human – that, in other words, will appear rational in an everyday sense.

Contrasting design with science will always be difficult to the extent that design and science are both complex activities. A simple comparison with deductive or inductive reasoning does not suffice because a logic-based computational system may not, in fact, seem so logical when organized in a complex way. Arguing that designers work in conditions of uncertainty or subjectivity does not pose a direct challenge to logical models. Direct challenges involve either the contention that cognitive processing does not utilize discrete, amodal symbols, or that there is a better level at which to develop explanations of problem solving altogether. Concerning the former, some have argued, for example, that cognition involves sketchy symbols (Goel 95), perceptual symbols (e.g. Barsalou 99) or no symbols at all (e.g. Port and van Gelder 95). Concerning the latter, others have argued that looking at representational form and other elemental features of cognition is premature and that, instead, we should be developing theories about socially and culturally situated activities using different theoretical concepts altogether (e.g. Greeno 98; Hutchins 95; Suchman 87).

By actually comparing what designers do with what scientists do when faced with real problems typical of their respective disciplines, one might, in any case, find that science is more like design than previously thought. Empirical studies of science, for example, show that scientists sometimes do exactly what Lawson (79) observed designers doing. Specifically, scientists sometimes construct provisional analog models which they then use to develop more abstract theories, particularly when operating in areas of uncertainty. Faraday, for example, designed and reflected upon a model of a mechanical system in an effort to develop his field theory of electricity and magnetism (Nersessian 99). In this regard, attempts to cordon off design from science may be premature. To use abduction, however, to describe either design or science requires that one drop the assumption that it can be explained as a singular type of reasoning (cf., Bylander et al. 91; Konolige 91; March 76) and that it instead be expanded to incorporate more complex and connected processes like productivity, analogy and visual reasoning (Thagard and Shelley 97).

3.2 Design as ill-defined problem solving

Reitman (65) identified ill-defined, or ill-structured, problems as those "whose definition includes one or more parameters, the values of which are left unspecified." His classic example was the problem of composing a fugue, which, in its simplest form, contains only the requirement that the result have the quality of fugueness. This quality, of course, is not defined in advance. Following Reitman, Newell and Simon (72) developed a problem space model of problem solving according to which ill-defined problems are those with variable problem spaces. Ill-defined problems, more specifically, include start states (e.g. given information and resources), operations that can be used to transform states (e.g. moving parts around, changing materials, etc.) and goals (e.g. criteria that can be used to judge states in the problem space), one or more of which are open to redefinition.

In a subsequent analysis, Simon (73), similar to Reitman, used large, decomposable design problems to exemplify ill-defined tasks. When a problem is gradually decomposed, the problem space is under continuous change, to the extent that the detailing of certain parts will be deferred until other parts have been completed. If one is designing a car, for example, it would not make sense to assemble all the necessary information at once into one problem space involving a fixed set of states. This is because decisions invariably propagate constraints, gradually narrowing the search space as the design proceeds. If one decides early on that the car will be a hatchback, for instance, there will be no reason to think at any point in the design process about the design of trunks.

In ill-defined problems, problem spaces gradually unpack themselves without clear foresight. Since, as Simon argues, decomposition is a result of an overabundance of information, "there is nothing other than the size of the knowledge base to distinguish [ill-structured problems] from [well-structured problems]" (73, p. 179). Cast in a slightly different way, any problem is ill structured if the problem solver learns more about it over time. In chess, for example, a game takes shape and is transformed as moves are carried out:

> "The ill-structuredness, by the usual criteria, of chess playing becomes fully evident when we consider the play of an entire game, and does not confine our view to just a single move. The move in the real game is distinguished from moves in dynamic analysis by its irrevocability – it has real consequences that cannot be undone, and that are frequently different from the consequences that were anticipated" (Simon 73, p. 186).

Given a general interpretation, ill-structuredness is not a feature unique to design problems despite its frequent use in describing design. In response, though, some have argued that Simon's framework does not adequately characterize just how ill-defined design problems are. Goldschmidt (97) for example, argues that defining an ill-defined problem is not simply a matter of retrieving information from a large database of relevant information. In particular, Goldschmidt argues that retrieval processes do not reflect "the 'messiness' of a search that we associate with the solving of ill-defined design problems" (p. 443). Others have also challenged Simon's view by arguing that design problems can be restructured or redefined at will, that constraints between components in a decomposition, for example, are "negotiable" (Goel 95) and that rules are "contingent and contextual" (Schon 88).

Having access to a large database of information may, however, still explain a "messy" solution process, one in which, for example, the subtlest of retrieval cues could trigger the retrieval of relevant information, resulting in an idiosyncratic structuring of the problem. Also, having access to a large database of information might be enough to explain the contingent or negotiable nature of constraints. Contingent constraints, for example, are not simply dropped in design but, rather, are redefined through substitution with available information. A designer might violate budget constraints not simply because he or she feels like it but rather because the budget constraint can be redefined as a cost-to-quality constraint, based in part on knowledge of how the client might feel about it. In this case, the highest-level (inviolable, but not well defined) constraint is satisfying the client.

Like non-design problems, all design problems also come with some fixed, well-defined constraints. A house, for example, must be able to withstand light breezes. One question then, is at what point will a problem cease to be ill-defined. Reitman (65) argues that whether a problem is ill-defined or not is really a matter of community consensus, insofar as conventional practices may dictate a set of closed attributes for what appears, on the face of it, to be a vague description of what is demanded:

> "To the extent that a problem situation evokes a high level of agreement over a specified community of problem solvers regarding the referents of attributes in

which it is given, the operations that are permitted, and the consequences of those operations, it may be termed unambiguous or well defined with respect to that community" (p. 151).

Simon (73), on the other hand, argues that all problems are ill defined to the extent that all are open to some restructuring. Consider, for example, a game where two players take turns picking numbers from one to nine, seeing who can first get three numbers that add up to fifteen (Simon 81). Although the problem is well-defined in all respects as it is given, it can be restructured if one notices that it is analogous to tic-tac-toe. Similar restructurings could potentially apply to design problems for which decompositions are specified by convention. In this case, whether a problem is ill-defined or not depends on the priorities and abilities of the problem solver, not by the type of problem one is given. Thus, turning a problem into an ill-defined one may be a creative step.

Figure 1: A two-player game in which players take turns selecting numbers from 1 to 9. The first person with three numbers adding up to 15 wins. By laying out the numbers in a grid with each row, column and diagonal totaling 15, the game is transformed into an isomorph of tic-tac-toe.

So how useful is it to characterize design problems as ill-defined? The definitions provided by Reitman and Simon do not appear sharp enough to distinguish design from other types of problems. There are a couple of ways those definitions can be clarified, however. One is by distinguishing problems that require constant structuring from problems that are only restructured discretely. The latter includes insight problems and other restructuring problems (such as the tic-tac-toe problem above) for which there are fixed solution criteria and transformation rules. Restructuring is possible (and sometimes even encouraged) with these problems, but within fixed boundaries. The former, on the other hand, includes design problems for which the space of potential restructurings is large enough to be effectively open. Problems of this sort can be restructured at any level of abstraction, even in the smallest details. The difference between design problems and other ill-structured problems is, in this case, analogous to the difference between a lump of clay and tinker-toys. Both can be restructured, but clay can be transformed at the molecular level, resulting in what seem to be almost fluid transformations. This distinction is consistent with Simon's (73) observation that although all problems are potentially open to re-definition or restructuring, some require that more time be spent structuring them than searching the space that results.

Another possibility is to argue that design problems involve an overabundance of information that happens to be social in nature. Information, in this case, is likely to vary from person to person. Awareness of a multiplicity of views may then lead not only to an endless expansion of information but also to ambiguity and conflicting views. Voss et al. (91), for example, have noted these features in ill-defined problems associated with the development of foreign policy. The next section considers the social realm as a distinct feature of design problems in more detail.

3.3 Design as wicked problem solving

In an influential article addressing general efforts to rationalize problem solving, Rittel and Webber argue that design and planning problems are "wicked" and that classical models of optimization do not apply (Rittel and Weber 73). Like Reitman and Simon, they argue that wicked problems are poorly defined, without clear start states, transformation rules or end states. However, they also point to the social embeddedness of planning and design problems, where values about good and bad solutions vary among stakeholders. Solutions to wicked problems change over time as they generate "waves of consequences over an extended—virtually unbounded—period of time." (p. 163). Every wicked problem can also be considered to be a symptom of another problem. Because of their connectedness to the environment and to other problems, solving a wicked problem may inadvertently exacerbate other problems. The moral responsibility of wicked problem solvers is thus seen as different than that of scientists. Whereas scientists must propose hypotheses that later are proved wrong, the solver of wicked problems seeks to improve some aspect of the world and hence has no right to be wrong, yet operates in an ambiguous social system. "The planner who works with open systems," Rittel and Webber observer, "is caught up in the ambiguity of their causal webs" (p. 167).

In Rittel and Weber's analysis of policy making, emphasis is placed on framing problems and how certain solutions and solution criteria are "naturally" preferred by some stakeholders. "The choice of explanation is arbitrary in the logical sense. In actuality, attitudinal criteria guide the choice. People choose those explanations which are most plausible to them. Somewhat but not much exaggerated, you might say that everybody picks that explanation of a discrepancy which fits his intentions best and which conforms to the action-prospects that are available to him" (73, p. 167). Rittel and Webber point, in particular, to the complexity of the social realm as a source of wickedness. There are too many potential side effects and casual chains to be able to say precisely what will result from social planning. Thus any attempt to circumscribe the potential outcome of planning will somehow be subjective and open to dispute.

Does wickedness, however, suggest that a design problem will be solved differently from other problems? Are wicked problems simply hard problems? To be sure, some problems which are not likely to be considered design problems share many of the features Rittel and Webber use to characterize wicked problems. Planning to jump across a ravine, for example, is a type of problem that cannot be solved either incrementally or by trial and error. It also cannot be solved with any certainty in the sense that there will always be a delay between committing to a plan and finding out whether it works. Furthermore, a number of factors, such as the stability of the soil on the other side of the ravine, may be ambiguous and require guesswork. Jumping across a ravine is not, of course, a socially embedded problem in most instances. How exactly then does social-embeddedness change the problem solving process?

Like ill-defined problems, the crux of the matter might be the overabundance of potentially relevant information. Jumping across a ravine for style points, for example, is more complex than jumping across a ravine for fun. Still, though, this may not be seen as altering the problem solving process in any radical way. Complexity simply demands simplification, like ill-structuredness demands structuring. Simon (81), for example, concludes that working in the social realm simply calls "for modesty and restraint in setting the design objectives and drastic simplification of the real-world situation in representing it for the purposes of the design process" (p. 163). Subjectivity, or social-embeddedness, in this case, may not be a distinct feature of design problems. Even if a problem involves a complex set of subjective values, in other words, a person holding those values may, in the end, act on them in a fundamentally rational way.

Another possibility is that some design problems, in contrast to non-design problems, are unique in that they specifically demand the consideration of "the essence of what the artificial may be in human experience" (Buchanan 92, p. 18). Voicing a similar sentiment, Winograd (96) draws parallels between software design and architecture, pointing out that when an architect designs a home or an office building, "the patterns of life for its inhabitants are being specified" (p. xvii). Design problems in the social realm may, in this sense, be significantly more complex than others. Simon (81), for example, points out that social embeddedness acts as a wild card: "The planners make their move and those who are affected by it then alter their own behavior to achieve their goals in the changed environment" (p. 178).

Architects, product designers and software designers may be taught to deal with the exponential complexity associated with the social realm directly, that is, by participatory interventions. However, other designers, particularly engineers, may work on artifacts that are ultimately hidden from consumers (e.g. a carburetor) or that are not meant to be appreciated directly (e.g. a waste treatment facility). Engineers may, of course, have opportunities to consider wide-ranging side effects in the social realm. An engineer, for example, who is concerned with the ecological impact of his or her work could consider how a particular design influences attitudes towards the environment and resource consumption in an extended context. Despite this possibility, however, engineers in such situations are more likely to focus on minimizing the consumption of resources and production of pollutants within the narrower context of a particular production process. This is simply because they are trained to deal with how artificial things work rather than how the social realm works.

3.4 Design as construction

Rather than reasoning solely from information derived from knowledge in long-term memory, designers construct concepts on the fly to help them further define and decompose problems (Ward 95). The constructive nature of design presents another difficulty for the argument that design can be characterized in terms of a series of logical processes. As such, it has been contrasted directly with information processing and search (Schon 88). At issue is the pervasive assumption that, in design, interrelations among well-structured subproblems can be ignored in ideal situations to the extent that a "good procedure will divide the task into components that are as nearly 'self contained' as possible" (Simon 73, p. 1). The emphasis on conceptual construction over information retrieval has become one of the ways in which design researchers distinguish design from other activities. It is also, however, one of the issues that may limit the degree to which research on design can be generalized across disciplines.

The idea of construction is straightforward: When a designer becomes "stuck in a problematic situation" that cannot be readily managed, he or she "may *construct* a new way of setting the problem – a new 'frame'" which is then imposed on the situation (Schon 83, p. 63). In one sense, constructing a frame, to use Schon's terminology, is like choosing a decomposition, except for the critical distinction between 'construct' and 'choose.' Consider, for example, early designs for flying machines. Initially, designers decomposed the problem following the structure of birds, unsuccessfully using wings to generate both lift and thrust (Isaak and Just 95). Eventually, these early attempts evolved into more successful designs that separated lift and thrust into different systems. The structure inherent in the later designs was not, however, an option early on. It came about only after a period of frustration, radically changing the way aerospace designers viewed flight from then on.

In addition to being associated with decompositions, frames also have a second connotation. Specifically, frames are also associated with perspectives taken by designers to help direct action and to organize unwieldy sets of information into coherent goals. Perspectives, in this sense, may influence the way a critical source of information is interpreted – information, for example, about the needs of end users. Perspectives may also, however, be more long-ranging, dealing, for example, with the treatment of design problems in general.

Details about the constructive aspects of design are lacking. Although Schon (83), identifies several ways designers might devise new situations when existing ones are causing trouble, much seems to hinge simply on critical reflection. Knowing the situation you are in, in other words, is assumed to illuminate opportunities to change the situation. As evident in some of Schon's observations, however, a designer's situation may also be transformed in more dynamic ways. Specifically, designers may generate mediating concepts that recast relationships between different modules as a design solution is being worked out. Designers, for example, may find opportunities to create features within an existing structure midway through a design, changing the direction of the decomposition at that point. A student designer observed by Schon, for example, transforms the overlapping areas of a series of building units that make up an elementary school into nooks for the students. This ultimately leads to a new decomposition that replaces one that was previously blocked.

The constructive aspect of design also comes to forefront when decompositions involve dependencies that work in many directions at once (see Akin's chapter in this volume). In design, dependencies between components are not limited to top-down or lateral constraints. Parts at the bottom of a decomposition, that is, may also influence a decomposition, or concept, as a whole. Consider, for example, Reitman's (65) observations of a composer working on a fugue. Initially, the composer breaks down the fugue into a series of stages, a move that is likely dictated by convention. In any case, top-down constraints initially influence the decomposition (the style of the fugue, for example, may dictate what type of components it will be broken into), while lateral constraints crop up between decomposed units (the end of the first part of a fugue, for example, must merge harmoniously with the beginning of the second part). Constraints, however, may also work backwards. Changing a part of a fugue locally could, for example, result in the consideration of a new style with new rules for how the parts themselves should be organized. Bottom-up influences on the overall structure of concepts are also evident in the work of the student designer observed by Schon. The spontaneous generation of nooks between school classrooms, for instance, might get one thinking about the way a sense of control and ownership influence learning and, in turn, reorient the design of a school from the top down.

At this point, one might ask whether construction is unique to design or, conversely, whether it applies to all types of design. Does construction, for example, apply to problems in science or other professions? Designers construct novel situations simply by reflecting on compositions that they have never thought about before. They assemble concepts triggered by the statement of a problem, for example, and look for novel organizational structures within that assembly. A physicist, or sociologist might also, of course, look for patterns in collected data. And although one might argue that designers, in contrast to scientists, deal with self-made or imaginary compositions, scientists similarly construct thought experiments and reflect on imaginary, analogous situations. Schon (83) ultimately backs the view that construction is not unique to design in that he uses clinical psychology as one of two primary examples of constructive practices.

At the same time, it is not clear that construction applies to all areas of design equally well. Some types of engineering design, for example, have been described as involving the composition of a fixed set of components according to a fixed set of compositional constraints (Chandrasekaran 90). Additionally, some kinds of engineering design may be more like repair, contrasting with those that involve developing complex social and technical goals that are constantly re-structured. In this case, like Simon's (81) treatment of ill-defined problems, even if the way compositions are searched can be reorganized conceptually, the components themselves and the underlying structural constraints remain fixed.

4 Summary

The dominant approach in design studies has been to claim design as a unique type of problem solving with unique features. Along these lines, a major goal of design studies has been to verify and catalogue these features, many of which deal with the structure of design problems and attributes of design behavior. It is hard, however, to maintain general, complete descriptions, simply because design is not a well-circumscribed activity. Simple definitions, moreover, seem to be limited to fictional situations much like general computational models of problem solving are limited to artificial worlds.

Nonetheless, the descriptions discussed above do point to the possibility of constructing a typology of design activities – one that might provide a conciliatory framework for researchers working in different disciplines. For example, although ill-definedness may not be unique to design problems, the complexity associated with the sorts of information that bear on design problems may demand unique structuring activities. Designers, for instance, may not only need to structure given information, they may also need to structure information resources. This, in turn, points to processes that involve the use of organizational schemas and perspectives to help make problems manageable. Ill-defined problems outside of design, like control problems (e.g. Funke 91), may, on the other hand, lack this feature in that they entail a given perspective; a control problem, that is, can at least be solved on the basis of a physically delimited set of information (e.g. information displayed on a monitor) even if variables and constraints are not explicit. Wickedness also appears to apply to a wide range of problems, even though it may not apply to all design-related activities. If a problem is wicked, dealing with the complexities of the social world, particularly reasoning about artifacts that are meaningful within a dynamic social context, may encourage activities like exploring and integrating multiple perspectives and speculating on issues in an ad hoc way. Wickedness also, of course, suggests a unique role for social interaction in solving problems. Construction, finally, points to other distinct features of a range of design-related activities, such as the role of analogical mapping, conceptual combination and other processes in the development of a decomposition.

Ill-definedness, wickedness and construction do not describe design in full. An overabundance of information, for example, may require analysis, idiosyncratic search, free-association, the use of organizational schemas and so forth, but none of these activities would ever be confused with design. However, it may not be important to construct a complete image of design in order to make progress in understanding design itself.

An alternative to starting with complete and distinct notions of design is to treat design problem solving as problem solving in general and focus individually on the reasoning processes that are involved. This does not mean taking a bottom-up cognitive approach in which basic cognitive parameters like memory and attention are clarified and integrated into more complex processes. The idea, rather, is to focus on types of reasoning – on things like analogy, coherence seeking, mental simulation, dynamic modeling, argumentation, decision making and so forth – as they contribute to the generation of new artifacts. Starting at a lower level of abstraction may be unproductive in that it fails to address the way such activities are situated in design (Greeno 98; Suchman 87). Analogical reasoning, for example, has been studied extensively in cognitive science, but the types of research conducted do not always seem to fit well with intuitive notions of design. While research in cognitive science has focused on the mapping of causal structures from one domain to another (Holyoak and Thagard 95; Gentner and Markman 97), analogies in design may be functionally unique. A software designer, for example, may imagine a computer-based communication system by simultaneously investigating how stock markets operate (Malone and Crowston 94), while an architect may design a city park by simultaneously exploring the social organization of sky scrapers (Koolhaas 78). In such cases, analogous domains are explored reflexively rather than simply used to map structures from one area to another. In some cases, analogical mappings in design may also hinge on the use of visual images (e.g. Casakin and Goldschmidt 99) rather than on propositionally encoded causal structures.

The strategy for investigating design proposed here – what one might call á la carte research – has two potential benefits. First, it puts design research in a position to reflexively contribute to cognitive science. By situating itself more centrally within cognitive science, research in design may be less likely to grow stagnant. Real discussions between design researchers and the broader cognitive science community may, in fact, refresh both. The second potential benefit is that it will help facilitate exchanges between researchers in different design disciplines. Rather than preface studies with blanket statements about design, which may or may not apply well to other disciplines, researchers would deal with aspects of real-world problem solving, relying on particular disciplines only for context.

References

Alexander C S, (1964), *Notes on the synthesis of form*, Harvard University Press, Cambridge, MA.

Asimow M, (1962), *Introduction to design*, Prentice-Hall, Englewood Cliffs, NJ.

Barsalou L W, (1999), Perceptual symbol systems, *Behavioral and Brain Sciences*, 22, 577-609.

Britannica, (2000), *Merriam-Webster's Collegiate Dictionary.*

Rambow R, and Bromme R, (1995), Implicit psychological concepts in architects' knowledge – How large is a large room? *Learning & Instruction*, 5, 337-355.

Buchanan R, (1992), Wicked Problems in Design Thinking, *Design Issues*, 8 (2), 5-21.

Bylander T, Allemang D, Tanner M C, and Josephson J R, (1991), The computational complexity of abduction, *Artificial Intelligence*, 49, 25-60.

Casakin H, and Goldschmidt G, (1999), Expertise and the use of visual analogy: Implications for design education, *Design Studies,* 20 (2), 153-175.

Chandrasekaran B, (1990), Design Problem Solving: A Task Analysis, *AI Magazine*, 11 (4), 59-71.

Chi M T H, Feltovich P, and Glaser R, (1981), Categorization and representation of physics problems by experts and novices, *Cognitive Science,* 5, 121-152.

Chi M T H, (1989), How inferences about novel domain-related concepts can be constrained by structural knowledge, *Merrill-Palmer Quarterly*, 36, 27-62.

Coyne R D, (1988), *Logic Models of Design*, Pitman, London.

Cross N, (1995), Progress toward an independent theory of design, unpublished presentation at the meeting of the Design Research Society, University College London.

Downing F, (1992), The role of place and event imagery in the art of design, *Journal of Architectural & Planning Research*, 9 (1), 64-80.

Frensch P A, and Buchner A, (1999), Domain-generality versus domain specificity in cognition, R J Sternberg ed, *The nature of cognition*, MIT Press, Cambridge, MA, 137-172.

Funke, (1991), Solving complex problems: Exploration and control of complex social systems, R J Sternberg, and P A Frensch Eds. *Complex problem solving: Principles and mechanisms*, Lawrence Erlbaum and Associates, Hillsdale, NJ.

Gentner D, and Markman A B, (1997), Structure mapping in analogy and similarity, *American Psychologist*, 52 (1), 45-56.

Gero J S, and McNeill T, (1998), An approach to the analysis of design protocols, *Design Studies*, 19 (1), 21-61.

Goel V, (1995), *Sketches of thought*, MIT Press, Cambridge, MA.

Goldschmidt G, (1997), Capturing indeterminism: Representation in the design problem space, *Design Studies*, 18 (4), 441-455.

Greeno J, (1998), The situativity of knowing, learning, and research, *American Psychologist*, 53, 5-26.

Hillier B, Musgrove J, and O'Sullivan P, (1972), Knowledge and Design, W J Mitchell Ed. *Environmental Design Research and Practice*, University of California Press, Los Angeles, CA.

Hubka V, and Eder W E, (1996), *Design science: introduction to needs, scope and organization of engineering design knowledge*, Springer, New York, NY.

Hutchins E, (1995), *Cognition in the Wild,* MIT Press, Cambridge, MA.

Isaak M I, and Just M A, (1995), Constraints on thinking in insight and invention, *The nature of insight*, R J Sternberg, and J E Davidson eds, Cambridge, MA, MIT Press.

Konolige K, (1991), Abduction versus closure in causal theories, *Artificial Intelligence*, 52, 255-72.

Koolhaas R, (1978), *Delirious New York: a retroactive manifesto for Manhattan*, Oxford University Press, New York, NY.

Lawson B R, (1979), Cognitive Strategies in Architectural Design, *Ergonomics,* 22 (1), 59-68.

Malone T, and Crowston K, (1994), The Interdisciplinary Study of Coordination, *ACM Computing Surveys,* 26 (1), 87-120.

March L, (1976), The logic of design and the question of value, L March Ed. *The Architecture of Form*, Cambridge University Press, New York, NY.

Mistree F, Smith W F, Bras B, Allen J K, and Muster D, (1990), Decision-based design: A contemporary paradigm for ship design, *Transactions, Society of Naval Architects and Marine Engineers*, 98, 565-597.

Morgenstern J, (1995), The fifty-nine story crisis: Engineer W. LeMessurier discovers structural flaws concerning wind resistance in his Citicorp tower in Manhattan, *The New Yorker*, 71, 45-53.

Murphy G L, and Medin D L, (1985), The role of theories in conceptual coherence, *Psychological Review*, 92 (3), 289-316.

Nersessian N J, (1999), Model-based reasoning in conceptual change, *Model-based reasoning in scientific discovery*, L Magnani, N J Nersessian, P Thagard eds, Kluwer, New York, NY.

Newell A, and Simon H A, (1972), *Human problem solving,* Prentice-Hall, Englewood Cliffs, NJ.

Port R F, and van Gelder T, Eds. (1995), *Mind as motion: Explorations in the dynamics of cognition*, MIT Press, Cambridge, MA.

Purcell T A, and Gero J S, (1996), Design and other types of fixation, *Design Studies,* 17 (4), 363-383.

Reitman W R, (1965), *Cognition and Thought*, Wiley, New York, NY.

Rittel H W J, and Weber M M, (1973), Dilemmas in a general theory of planning, *Policy Science,* 4, 155-169.

Roozenburg N F M, (1992), On the logic of innovative design, N Cross, K Dorst, and N Roozenburg Eds. *Research in Design Thinking*, Delft University Press, Delft, 127-38.

Roozenburg N F M, and Cross N G, (1991), Models of the design process: integrating across the disciplines, *Design Studies,* 12 (4), 215-221.

Schon D A, (1983), *The Reflective Practitioner: How professionals think in action*, Basic Books, New York, NY.

Schon D A, (1988), Designing: Rules types and worlds, *Design Studies,* 9 (3), 181-190.

Schon D, (1992), Teaching and learning as a design transaction, N Cross, K Dorst, and N Roozenburg Eds. *Research in design thinking*, Delft University Press, Delft, 21-36.

Simon H A, (1973), The Structure of Ill Structured Problems, *Artificial Intelligence,* 4, 181-201.

Simon, (1981), *Sciences of the Artificial*, MIT Press, Cambridge, MA.

Suchman L, (1987), *Plans and situated actions: The problem of human-machine communication*, Cambridge University Press, New York, NY.

Thagard P, and Shelley C P, (1997), Abductive reasoning: Logic, visual thinking, and coherence, M-L Dalla Chiara, K Doets, D Mundici, and J van Benthem Eds. *Logic and scientific methods*, Kluwer, Dordrecht, 413-427.

Vera A H, and Simon H A, (1993), Situated action: A symbolic interpretation, *Cognitive Science*, 17, 7-48.

Voss J F, Wolfe C R, Lawrence J A, and Engle R A, (1991), From Representation to Decision: An analysis of problem solving in international relations, R J Sternberg, and P A Frensch Eds. *Complex problem solving: Principles and mechanisms*, Lawrence Erlbaum and Associates, Hillsdale, NJ.

Ward T B, (1995), What's old about new ideas, *The Creative Cognition Approach*, S M Smith, T B Ward, and R A Finke Eds. MIT Press, Cambridge, MA.

Winograd T, (1996), Introduction, *Bringing design to software*, T Winograd, J Bennett, L De Young, B Hartfield Eds. Addison-Wesley, New York, NY.

Chapter 8

New Directions in Design Cognition: Studies of Representation and Recall

Charles Eastman
Design Computing
Colleges of Architecture and Computing
Georgia Institute of Technology
Atlanta GA

Abstract

The purpose of this chapter is to pose some new questions for design cognition and outline some paths for studying them. After a short review of design cognition's current status, I pose some issues that seem important to understand if design education is to be improved. Two broad areas are addressed:

(1) Learning new representations, their internalization and automatization, as a foundation for developing design expertise;

(2) Learning and recall used to structure a design context, framing the design task and retrieving relevant design concepts.

I survey recent work in the psychology of mental imaging and knowledge representation as a basis for addressing these issues. Studies in these two areas are reviewed and new ones proposed and the benefits of the new directions outlined. Two design domains are considered, architecture and mechanical engineering.

1 Introduction

We now have over 30 years of work in the area of design cognition. I use the term "design cognition" to refer to the study of human information processing in design, using different theoretical and empirical paradigms. Design cognition has become a defined field with a high quality journal, several conference series and some major references. The intention of this volume has been to take stock of the current status of design cognition, especially with regard to knowledge useful in improving design education. This chapter poses some new questions for design cognition and outlines some paths for studying them.

Below, I review the major results of work to date and the methods used. Other chapters of this volume survey these issues in more depth. The rest of this chapter is organized as follows. After the review, this first section ends by posing a set of issues of importance to design cognition, especially design education, that have not yet been well addressed. These issues form the framework for the succeeding sections. Section 2 reviews work in mental imaging and knowledge representation, which are proposed as foundations for new studies for better understanding the education of designers. This section reviews and interprets this literature, without making the strong connections to designing; these are made in succeeding sections. Section 3 applies to design what is known about learning and using external representations and their internalization and mental use. Section 4 reviews studies of how designers structure a design context, recalling relevant criteria and features. It builds upon recent work in knowledge representation, memory and recall. Methods for studying the issues are outlined as we go.

1.1 A Review of Design as Ill-defined Problem solving

The emergence of cognitive psychology in the 1960s offered both a conceptual paradigm for describing design, as well as a method for studying what is mostly an invisible, mental activity. The initial efforts drew on earlier work in problem solving in cognitive science and artificial intelligence (these two fields initially had common roots (Feigenbaum and Feldman 63)). Design was initially studied as a type of problem solving (Newell 69), as a search of a space of possible solutions for the best or a "satisficing" solution, in an approach similar to studies of chess, crypto-arithmetic, and puzzle solving (Eastman 70).

Design was quickly recognized as different than other forms of problem solving. The structure of the problem is usually not given and the criteria applied are at best only given abstractly, seldom in terms that can be directly applied. The terms "ill-structured" and "ill-defined" were introduced to signify that design tasks required definition of the problem space and also the criteria applied to candidate solutions (Reitman 64; Simon 73; Akin 86). Design researchers have studied various sources of information (Eastman 69; Visser 96) including access to relevant physical artifacts (Harrison and Minneman 96).

It was also recognized that large design tasks are decomposed and organized into multiple levels of detail and different functions (Baya and Leifer 96; Ullman, Herling and Sinton 96). Studies were undertaken regarding how mental resources were allocated within this structure (Gunther, Frankenburger and Auer 96), how these different tasks were organized within the whole process and how designers iterated between the different sub-tasks (Purcell, Gero et al. 96; Akin and Lin 96) and how designers manage time (Baykan 96; Radcliffe 96). Others have considered the social roles people play within design teams (Cross and Cross 96; Brereton, Cannon et al. 96)

Parallel to the study of design behavior has been the analysis of the structure of design problems. Better understanding of the structure of some design spaces may allow further insight into the processes followed by designers. One popular structuring of design information has been to distinguish between structural, behavioral and functional information (SBF) (Takeda, Tomiyama et al. 96; Goel, Gomez et al. 96). *Structure* is the form—materials and geometry—of the design solution, *functions* are the general objectives to be realized, and *behavior* is the measurable performances into which the functions are translated.

A general insight was that traditional search within a narrow design space was undertaken only occasionally, in response to specific situations. More often components, features or attributes were identified from the general goals (Akin 78). That is, design (at least the early conceptual parts of design frequently studied) was more involved with defining the solution space than the search of specific solution points within it. However, later stages, particularly those in engineering design, focused on decisions regarding solution points (Ullman, Herling and Sinton 96).

1.2 Protocol Analysis

The primary empirical method for studying design introduced at the beginning of cognitive psychology was *protocol analysis*. Protocol analysis involves giving small but realistic design tasks to subjects and monitoring their behavior. Data was collected using video, plus any drawings produced. Design thinking is induced from the behavior captured from the protocol, including verbalizations, drawing and gestures. The methods and formal assumptions of protocol analysis were laid out in Ericsson and Simon (84). They presented it as both an empirical method for studying behavior and as a theoretical approach, based on the problem solving paradigm and mental processing using symbolic knowledge representations (to be discussed later).

1.3 Critique

There has been a steady stream of criticism regarding both protocol analysis as a method of empirical method for studying design and ill-defined problem solving as a paradigm for design. Protocol analysis relies heavily on the information provided in external representations, both verbally and in drawings and notes. There has been criticism of the distortions introduced when verbalization is forced (Davies 95; Lloyd, Lawson and Scott 96; Baya and Leifer 96). Protocol analysis, as described in Ericsson and Simon, does not address well the differences between internal and external representations (Chi 97). Another criticism is the gap between the levels of description offered within most studies based on protocol analyses and a designer's perception of that he or she is doing (Dorst 97, c 1). Variations and refinements of the task definition and for analyzing the behavior have been made, leading to a family of related empirical/formal methods of study (Chi 97; Crismond 97) (see Craig's chapter in this volume). Others have approached design through various forms of case study (Krauss and Meyer 70; Candy and Edmonds 96).

Many groups have also criticized the ill-defined problem solving design paradigm. Winograd and Flores have argued that it is ill-suited for studying design, noting that designers are part of the situation in which they act. Designers mentally construct their view of the situation as well as the actions taken within it (Winograd and Flores 87). They and others refer to the Heideggerian notions of "throwness" and "ready-to-hand" (Heidegger 62). With these concepts, Heidegger means to articulate the embeddedness of designers' actions in the real world. Partially from this perspective, some studies have approached design from different paradigms than ill-defined problem solving. Some have approached it as a socially organized set of actors (Brereton, Cannon et al. 96) and also based on a contextual situatedness (Cross and Cross 96). Others have taken a more phenomenological perspective, arguing that the designer mentally constructs a design world (Trousse and Christiaans 96; Schon 88) beyond the entities, attributes and relations usually employed to define a problem state space. This included mental simulations of use, expected contextual changes over time, and other issues that go far beyond the defining of parameters of a state space (Schon 92; Dorst 97).

The information gained from the duality of protocol analysis and ill-defined problem solving have been primarily normative, identifying common aspects of design as an activity. The motivation of much of this work is to articulate the distinctions of design cognition from other forms of problem solving. It has also sought to develop a general taxonomy of design cognition activities. While there has been little work clearly distinguishing processes supporting good design from less good, it has allowed identification, along with more prescriptive approaches (Hubka and Eder 88; Suh 90), of some of the constituent activities used by experienced designers. These include definition of the context of the design, examination of the design from multiple perspectives, generation of multiple alternatives, formulating critical performances, etc. These results have been used to identify differences between beginning and more advanced designers (Atman, this volume) and to assess progress in design education (Atman, Chimka et al. 99).

While research in design cognition has been successful in identifying *what* designers do, it has been less so in identifying *how* they do it. How do experienced designers frame design problems with regard to their context? We know that experienced designers are different from beginners; but what were the learning experiences that lead to the differences? How are multiple perspectives generated? How are performance criteria identified that are not given? What mental capabilities are used to generate alternative design concepts or features? These are new questions that could not be framed or articulated ten years ago. Their articulation is one of the contributions of recent work in design cognition.

A second area where protocol studies have not shed much light involves design representations. Design and engineering involve special representations that are central to their fields. Architecture emphasizes plans, sections and 3D models (manual or computer generated) and sketching. Engineering emphasizes mathematical modeling, especially differential equations, and various forms of diagram, such as circuit, network or kinematics. While these representations are taught and used throughout a university education, questions exist regarding how they are learned, applied and integrated with other knowledge. For example, there are concerns that architecture students select external representations based on their visual effect rather than what they communicate (Johnson 97). In engineering, students are taught to solve analysis problems, but when given an unstructured design task, often have difficulty relating their analysis knowledge to the task at hand (Emkin 95). How these representations are used, and especially how they are related to other mental constructs, seem central in understanding effective design.

It is generally assumed that the manipulation of external representations is only a partial depiction of what is going on mentally. The "real" structure of the design task resides in the designer's head (Lawson 97, c 8; Dorst 97, c 2; Downing 87). Cognitive models of thinking assume that external representations are auxiliary structures enhancing but not necessarily isomorphic to the designer's own knowledge structure. In design education, it should be the designer's mental representation of the context and current design that are the main concerns. These types of issues deal with mental processes usually not articulated in the data collected in a design protocol study. While protocol studies have allowed identification of the different external representations used (Ballay 87; Goel 95), there have been no traditional protocol studies to my knowledge that have shed light into the internal representations used by designers.

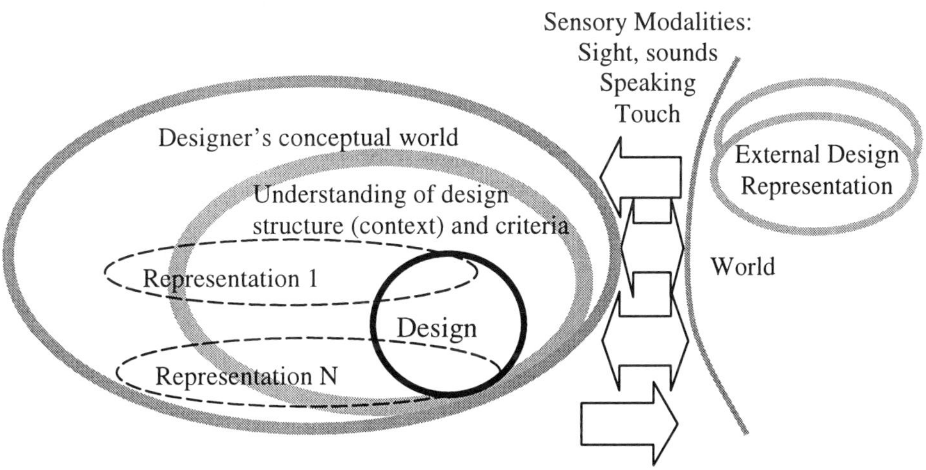

Figure 1: A designer's conception of a design and its context is built up over time, using information from the designer's already gained knowledge and experience, and from external sources of information

1.4 Recasting of the Issues

The above design issues can be recast in a manner different from ill-defined problem solving. By doing so, we can identify a different set of issues of importance to design, defined in a way allowing them to be fruitfully studied. The concepts of interest are partially characterized in Figure 1.

The central area of concern is the designer's conception of the design world, that is, of the design and its context. To develop this design world, a designer has multiple sources of information: the external world of objects and observation (example designs, the context of other designs), external sources of encoded information (such as books, the Web, drawings), and the internal recollection of previous experience and learning. The external sources provide information in different encodings. Books are encoded in words (in various colloquial or formal vocabularies) and mathematical formulas (in different notations such as sets, algebra or calculus), the physical world in the form of images and possibly tactile, motor and auditory memories. Information sources include drawings in different forms, such as floorplan, section, orthographic mechanical drawing, electrical, kinematics and other diagrams. From these sources, the designer has learned and mentally encoded a rich structure of information. Some of the designer's knowledge is *factual* (sizes, material limits), some *informal* (the processes of construction or manufacturing) and some is *procedural* (how to use a CAD system, how to use the Web) and some is tacit and *experiential* (what your car engine sounds like when running well or poorly).

Why do some beginning designers ignore their own experience when dealing with a design task? Why do others ignore what they have learned in college courses when they encounter a real problem? Why does verbalization seem to have a strong influence in developing design skills? These issues focus on how design information is carried in people's heads, how it is recalled, structured, and particularly, how it is manipulated. Design expertise seems to draw upon knowledge in different representations and benefits from reasoning across the different representations. Knowledge representation, recall, use and structuring are central issues not only for design but for all of cognitive psychology (Glass and Holyoak 86, c 1).

A central task of design is defining its context[1]. This includes both the *external context*, ranging from the physical, social and cultural setting of a building to the mechanical,

[1] Here, I have switched terminology used to describe design. Ill-defined problem solving uses a terminology that is inadequate and unevocative of the rich structure encountered in most design situations. From here on, I use the term *design context* to identify the outer environment involving system, physical, social, cultural, environmental contexts. Design context identifies the functions in the SBF terminology and their translation into desired behaviors. *Design response* is the intervention made by designers and corresponds to the structure in SBF terminology. These are similar to Simon's notion of inside and outside environment (69:c 1) and Alexander's notions of setting and of design as

control and human interfaces of a mechanism, to the fabrication technologies, operating capabilities and resources available for making and operating the product. The *designed response* responds to the context by changing or adding something into it, in realization of some goals or intentions. Defining the context and the repertoire of materials and methods that make up the designed response are core issues in effective design. How do expert designers conceptualize the design context, and the designed response? How is relevant design information identified and applied? All design fields assume that designers rely on a wider base of information than is explicitly taught; experienced designers draw upon both formally learned knowledge and also information continuously learned experientially i.e., "in the field". Design education, then, ought to include a useful scaffold for building such an experience base, well integrated with analytical and other technical knowledge acquired in school. Strategies for expanding and using such an experience-base should be considered a fundamental part of design education.

As potentially relevant information is recalled, it is in some sense integrated. During and after integration, a design is represented externally in a variety of representations: drawings (of various types), written specifications, numerical properties and their units of measurement. What is the relation of these external representations and the knowledge carried in the designer's head? Most studies of design have focused on the external representations (in part because they are accessible.) While it is intuitively clear that external representations are important, how do they relate to the designer's internal representations? Are external representations simply records of mentally determined changes? Or are they used to manipulate the design in a more generate-and-test manner with mental representations focusing on the criteria to apply and the procedural knowledge used to manipulate designs? Vinod Goel and a few others have begun to probe the general role of external representations in design (Ballay 87; Goel 95; Ulusoy 99) but we know little about how the external representations used by designers relate to their internal representation. Schon's investigations offer provocative insights (83). Goldshmidt has studied the role of sketching in preliminary design (91, 94) and suggested some relations between sketching and mental representations.

Some design knowledge is procedural. It includes the overall procedures needed to complete a design in some domain, to strategies for coordinating design actions within a team, to overall strategies for generating results emphasizing particular characteristics, such a high performance or economical operation, durability, or innovativeness. In design education, these procedures are sometimes reviewed explicitly. Equally often they are

---

correcting "misfits" (64). Together these two contexts are my focus in considering "design worlds". All human social activities additionally takes place in a cultural, organizational and social setting that also is part of a design world, and important for effective action. I ignore these last contexts here, focusing on the technical aspects of design.

presented tacitly in the context of a studio or project course. It has been this procedural knowledge that has been the focus of most design research to date.

Recent advances in psychology provide a background and conceptual framework to study many of the detailed questions raised above. The next section presents an overview of some of the work in mental imagery, knowledge representation and related areas that seem to provide a useful basis for studying and understanding these issues.

2 Mental Imagery and the Structure of Knowledge

The mental representation of knowledge has been an issue of serious study since Plato's discussion of the theory of forms in the *Meno* and throughout the history of psychology (Maier 31). As psychology developed in the early twentieth century, mental imagery seemed personal and without any measurable trace; it was easily disputed and seemed not accessible through empirical studies. Most behavioral psychologists, who were trying to make psychology a serious science, held imagery in disrepute. Later, the beginning work in cognitive psychology also did not accept the notion of mental imagery as a form of reasoning. Their early efforts focused on problem solving tasks that could be easily represented symbolically, such as crypto-arithmetic and chess. Newell and Simon argued for a node-and-link structure of semantic memory (72, c 14), (but later broadened their view, see (Simon 76)). Pylyshyn (73) argued that there were no "little men" in one's head to look at images and argued strongly for a symbolic, propositional structure.

Yet Paivio (71) found that learning ability was influenced by the ability to mentally visualize the information referents. Cooper and Shepard (73) showed that the response time required to match 2D and 3D mental images was proportional to a degrees of rotation required to the source and target images. That is, measurable performance carrying out some operation could distinguish whether people used mental images in some task rather than a node-and-link or other symbolic structure. The initial mental operations associated with imagery were applicable to perceptual images, such as rotation of a form, scaling it or measuring distances between forms. A variety of experiments indicated that these capabilities were easily carried out mentally and not coherently explained using symbolic representations. Developing tests for these types of functions, attempting to distinguish mental representations by their functional capabilities, led to many debates and continued refinement of experimental technique (Kosslyn 96, c 2).

Over the last fifteen years, a wide variety of operations on mental images have been studied. Among these are interesting capabilities for mentally operating on 3D shapes, including the integration of multiple orthographic views into a 3D shape and then recognizing new views of the shapes (Cooper 89), mental folding of a flattened shape and judging joined edges and other properties (Shepard and Feng 72), emergence of new shapes based on mentally composing known or easily understood 3D shapes (Finke 90). Shepp

(89) has begun to identify how visual features of shapes are extracted from an image and the properties of the features and their articulation during an infant's maturation. There is no straightforward or coherent explanation of these activities based on symbolic processes.

Evidence suggests there are varying types of mental images, ranging from 2D images (possible derived from 3D models) to detailed 3D models, to more abstract topological relations of connectivity (Denis 96). Johnson-Laird specifically argues there are at least two levels of imagery, (though his arguments seem to argue there are three levels of model) (Johnson-Laird 96). One kind of image is a *mental model*: an abstracted spatial form that can carry different kinds of relationships: topological, directional, etc. A mental model roughly corresponds to a diagram. A *3D model* is a complete shape in 3D from which 2D views may be projected. Last, there are *2D images*, which resemble pixel images with a fixed size and resolution. Fodor proposes a large, indefinite number of mental image types (75). Most researchers take the position that a mental image has a limited amount of information fully. Complex images or 3D models are not retrieved whole from long-term memory; rather they are constructed from a collection of imagery information about the object of interest (de Vega, Inton-Peterson et al. 96; Kosslyn 80). This suggests that it can be constructed in different ways, depending upon the need. These different functional capabilities have been only partially explored.

Recent neuroscience evidence also supports mental imagery. The issues being debated can be cast in terms of the functioning of emsembles of neurons (Kosslyn 96, p. 4). Using positron emission tomography (PET) scans, the visual field was found to have a specific mapping to adjacent neural regions (Fox, Mintun et al. 86). Moreover, soldiers with brain damage sometimes lost functioning in part of their visual field, and also the ability to form mental images in that part of the field, suggesting that stimulation of the region was not an epi-phenomological event, as argued by some psychologists, but was a functional one (Fareh, Soso and Dasheiff 92). These studies verify the sharing of some processing pathways by both perception and imaging (Kosslyn 96, c 11).

This debate evolved so that it was no longer about whether people experienced mental imagery; it was acknowledged by most researchers to be a personal phenomenon. The question was whether imagery could be used for reasoning. Similarly, the imagery adherents were not taking the position that all thinking is based on imagery; language processing and mathematical processing was readily accepted (vision tasks usually had to be given to people in words.) Rather, the question was how these internal representations are used, the capabilities associated with them and the role of referents and cross referents (Kosslyn 96, c 1). Based on the kinds of functional capabilities reported above, it is now generally accepted that people can do some reasoning using mental imagery.

All of this work acknowledges what I suppose all architects and most engineers have always known; people can integrate multiple drawings and interpret the forms and spaces they define. They can mentally simulate various activities in as-yet-unbuilt spaces and use such simulations to assess the space. However, researchers have also begun to map out the general capabilities and limits of these mental skills. There also are a variety of reports regarding individual differences in the ability to form and process mental images (Kosslyn, Brunn et al. 84; Denis 96) and different strategies to deal with mental tasks, some involving imagery and some not.

What is the basis for these individual differences? Are they innate, genetically determined faculties or are they developed by exercising them? All people seem to have some level of proficiency in mental imagery. There have been a small number of studies that indicate that mental imagery skills can be easily improved through training (Wallace and Hofelich 92). In an attempt to get to a finer grained neurological level of imagery processes, Kosslyn has broken it into nine component functional capabilities (96, c 11). They include: an attention window, low-level pattern activation and matching, exemplar activation and matching, spatial relation encoding, categorical encoding and so forth. He and others have begun to generate evidence for individual variation in performance for each of these different mechanisms, rather then assuming mental imaging to be a single capability (Kosslyn 94, p. 395-407).

The imagery debate has been carried out somewhat separately from other sensory modes. A significant body of literature exists on the cognition of sound recognition and interpretation, body movements and taste. They provide strong evidence that learning, recall and mental reasoning occurs normally in each of these sensory-related modalities. For example, people are able to dynamically compose muscle actions (i.e., in writing or playing a sport) (Lindemann and Wright 98; Viviani and Cenzato 85) and they can interpret unique compositions of sounds (as in listening to music or in interpreting what a person is doing in the next room from the sounds) (Bregman 95). Learning and reasoning regarding taste is a capability from infancy on (Harris 97). Thus the more general position is that there is some level of reasoning capability in most or all of our sensory modalities. The current issues involve the roles played by the sensory modal memory and knowledge in conjunction with the mental visualization, propositional processing and other forms of conscious reasoning. For us, an issue is if and how multi-modal information is used in design.

2.1 The Structure of Knowledge

If we store and later recall experience in each of our sensory modalities, and also have large amounts of symbolic information that is learned, how are these different types of knowledge organized, related and operated on? This question poses a current area of intense paradigm-building and research. Yet the general outlines, in relation to our interest in the structures used to carry design knowledge, can be defined. Most of the interpretation presented here draws heavily from Barsalou (93) and Kossyln (80, 96).

One starting point is the shared processing of visual and other sensory perceptions and processing of sensory memories. Many theories of cognition are strongly related to perception (Langaker 86; Talmy 88). Perception is not the photographic capture of an environment. It rather relies on focusing on aspects of the whole phenomenal situation, whether the aspects are visual, kinesthetic or auditory, recording mostly those aspects attended to (Kossyln 96, c 3; Barsalou 93). Selective attention largely determines what is encoded in long term memory (Craik and Lockhart 72), distinguishing vague recollections from sharp memories.

Encoded perceptions are thought to be composed according to what Crowder (93) called *procedural theories* that organize the memory of an experience in similar neural units to those that processed the original experience. Barsalou calls this knowledge structure a *structural description* that is built up relationally and hierarchically (93). Perceptually, people see or hear a pattern, match it with an identifiable structure, then fill in the details in succeeding moments based on the assumed structure (Kosslyn 96, c 6). These procedural[2] knowledge-structures are built up by "focusing strategic processing to the perceptual experience and extracting them as individual components" or chunks. Temporal or spatial relations between different inputs provide an initial automatic level of integration.

This does not mean that pixilated images are never stored, but rather that they require conscious attentional focus, and upon recall are part of a larger structure. Such an image has a limited level of detail and is located relatively within a spatial configuration (Kosslyn 96, c 10). Singular mental images are of a detail and scale to correspond closely to the image captured within the retina. Multiple images may be composed and scanned over to identify relations (at a lower resolution). Similarly, sounds, strong feelings, smells and kinesthetic behaviors are similarly stored, if they are originally attended to. These perceptual structures may represent multiple states over time, corresponding to what Langacker called a "cognitive grammars" (86). These perceptual structures may carry a mixture of perceptual modes, including aspects of sounds, images, smell, and feelings, resulting in a potentially rich remembrance of an event. As a perceptual structure becomes familiar and retained in long-term memory, it can receive incremental elaboration. Thus a recognized face may gain associated tones of voice, a collection of expressions and the moods experienced with them, and so forth. Whether these structures are the same or different from the more logically based propositional features is an area of current contention.

Experiences and other knowledge stored in memory are made more accessible if they have differentiatable cues. Word descriptions or names are the obvious example cues. Words, encoded as sounds or images, have an arbitrary but culturally-defined relation with an

---

[2] Here, I am using procedural in Crowder's (Crowder 93) sense, that memory is associated with the senses that originally received it.

appropriate perceptual structure. For example, the word "dog" or alternatively "chien" is related in ones experience with a class of house mammals that bark. In this way words and other symbols give us access to the properties of our perceptions. Symbols can be associated with experience in this way, to form a propositional structure for communicating experiences, from our past or as projections of the future. These propositional structures provide us with ways to access, and make temporal, numerical or logical inferences on our experiences—such as estimating the mileage from the time taken for a trip that included several stops. Thinking is not made up of logical operations on some isolated symbols alone.

In experience, Barsalou (93) suggests that propositional structures are developed secondarily from the procedural, experientially based ones. This explains why we are sometimes unable to initially express our experiences. Our experiences may initially have no verbal referent or stored interpretation. The first time someone describes an experience, it is a type of translation. In making the translation, linkages are built that enrich and elaborate the initial structure.

Later, upon recall of some event or concept, we recall only small portions at a time of what we know about it. That is, we may know a great deal about some concept or category, i.e., "dog", but we tend to recall only seemingly relevant sub-parts (Barsalou 93). If we ask the same question about a design of a doghouse next week as we did today, we are not likely to get the same answers.

So far I have emphasized the encoding of experiential information and treated symbolic information as secondary. But huge amounts of our knowledge are not directly experienced, but gained through external symbolic representations.

2.2 External and Internal Representations

By the time students reach the university, they are expected to have good skills in learning information and reasoning in several symbolic representations. They are expected to be able, at some level, to process them mentally, for example to make inferences, to identify inconsistencies, or to do mappings between them, such as between words and math or describing in words sensory experiences. These symbolic representations include written and spoken language and various forms of high school mathematics.

How were these external representations learned? Consider for example how we all learned to write. In primary school we learned to recognize and write the alphabet. We also learned the hand-eye coordination to write them. We practiced writing by filling pages of As, Bs, Cs etc., paying attention to each letter form and our mental tracing of them, until the muscle actions became accurate, smooth and automatic. Study of learning muscle movements for acts like writing show that such movements are organized hierarchically, with low level muscle movements being different, for example, for writing on paper versus a chalkboard, while the high level controls are the same (Lindemann and Wright 98). Successful students after several years are able to write fluently, meaning that the processes have become *automatic*, transparent and relatively effortless. They are not aware of the complex muscle and mental process of writing, but can focus directly on their ideas and phrasing. Similarly, with practice, typing can become automatic, resulting in the direct expression of ones thoughts on paper. Learning to read develops in a similar way. We began by learning to identify the alphabet and also whole words. We then put them together, drawing upon our existing knowledge of spoken language. With much practice, the process of reading becomes automated, allowing the meaning of the words to be received directly through the transparent interface of visually interpreted words and sentences. The processes of learning to read (Massaro 98) and to write (Gregg and Steinberg 80, Ericsson and Charness 94) have been well studied and documented.

Written language and math are taught as external representations, using paper and pencil. Through practice, two things take place: skills in reading and interpreting the representations are developed to the point of automatization, so these activities can be done without conscious effort; secondly, they are internalized, allowing some processes initially learned externally to be carried out mentally. That is, the learned math representation allows us to solve simple problems in our heads. Learning to read and write allows us to phrase and compose words in our heads, often used for mentally planning what to write (as this author is doing sitting at his word processor). The point is that learned external representations become internalized and become part of the mental representation capabilities available for cognitive processing. They become an additional intrinsic part of our reasoning capabilities. The ability to learn new representations and to internalize them is not limited to written language and high school math. It can also apply to a variety of other special representations. The range of representations potentially available for thinking is open-ended.

This view of cognition emphasizes that we have many different ways to internally represent information. People may attend to different aspects of the same event and gain different information from it. We often have multiple alternative representations and solution methods for dealing with a single problem. Current understanding allows us to appreciate several phenomena; explaining an emotion requires translation from the original modality to spoken language. Explaining a design action may also require translation and be difficult, if the reasoning for it was not verbally based. Different experiences and learned information need not be consistent nor fully integrated. Different people may internally represent the same experience differently, and interpret it differently. Short-term attention to recalled information (what might traditionally be called reflection) may identify conflicts and possibly resolve them. Reflection can also identify new relations between stored information, developing inferences. Using mental imagery, we can often simulate experiences or events only conceived mentally, imagining how we might walk through a space or view a scene. This view suggests the richness, power and great variation of uses available in human cognitive abilities.

This rough, abbreviated overview of recent research has attempted to summarize what is known of how humans mentally represent and structure knowledge and especially some of the capabilities regarding mental imaging. It identifies at a crude level the cognitive apparatus with which design students are thought to function. In the next two sections, I reframe the questions raised at the end of Section One according to this work and to pose them in a form that can be studied and potentially answered. First, I consider the teaching of design representations and whether they should be considered external and/or internal representations. I also address some of the roles of mental imagery in design.

3 Design Representations

All design students are taught new specialized representations at the university, as a foundation of their design education. These add to the representations learned earlier. The new representations broadly distinguish different types of design fields and serve as a base for more advanced learning.

Calculus and other forms of mathematics are among these advanced representations, taught as a foundation for advanced topics in engineering. Students are expected to able to represent problems presented in words in the proper mathematical form and also to interpret real world cases. They are expected to be able to take the necessary measures and assign appropriate values to the parameters of the math models. Example operations for reducing the mathematical representations and solving them are given in class; homework assignments give students experience in applying similar solution methods. Students are expected to be able to plan sequences of operations to be executed using paper and pencil (or a math package such as Matlab®).

While mental planning of operations is required to be minimally competent in advanced math courses, experts in calculus are able to do more, to represent problems and carry out operations on them in their head, without paper and pencil. One need only listen in college hallways after a calculus or physics test to hear students discussing their process for solving the most recent problem. These discussions are held without paper and pencil, with the participants carrying the formulations in their heads and describing simple operations. It is clear that some students, possibly most, are able to internalize the representation of calculus and to operate on it mentally. Is this the norm or an exception? Is the mental exercise of solving calculus problems in ones head educationally of value? To what degree does learning calculus to the point of being able to represent and operate on it mentally significantly benefit student engineers?

Mental representation is one possible competence threshold. Another is familiarity with the symbol structure of calculus to the point of being able to interpret it automatically. Like written words, the interpretation can be learned to the point of automatization. Are there significant performance differences between students that know calculus to the point of interpreting it automatically and those that do not? An indicator of automated interpretation would be fast response time in processing simple calculus task. Do such measures correlate with scores in more advance courses? To what degree is internalizing the representation a measure of competence, in comparison to developing strong strategies for applying and manipulating integral and differential equations? Math competence is currently judged according to aggregate scores in tests. Such test scores ignore time and effort, teamwork and other compensating tactics. More detailed analyses, identifying specific representation-specific cognitive skills and capabilities could lead to refinements of how calculus was taught, and provide better diagnostics in response to student problems.

In architecture, the main representations learned are architectural drawings—plans, sections, elevations—and 3D models. The drawings may be generated manually or on a computer; the models may be physical or computer-based. Students are assumed to quickly gain the capability to interpret architectural drawings and to interpret the 2D layout described. They are also taught, to varying degrees of proficiency, to generate architectural drawings of plans, elevations and sections. 2D drawing is taught within the context of 3D representations, so that the relation between 2D and 3D is given as part of the course context; the drawing process involves mentally mapping from 3D models or physical shapes to plans, sections and elevations. 2D floorplans and sections are important abstractions of a building of immense use in spatially understanding many kinds of circulation, space allocation, construction and other problems. 3D forms and spaces provide another more perceptual representation of a building useful for addressing other issues, including lighting, environmental comfort, and acoustics and appearance (Hewitt 85). Students are expected to gain a set of skills with regard to external representations: to be able to select appropriate representations, represent existing or proposed buildings within them, to be able to map between different representations in 2D and 3D. They are expected to be able to make inferences regarding a building's function, carry out mental simulations of activities, assess perspective views and other complex tasks. All of the later capabilities are built upon the ability to "read" and "write" architectural drawings with fluency[3].

Up to the last decade, architectural drawing has been a basic course in architecture curricula. Today, many schools of architecture have eliminated such courses, assuming that it can be picked up within the context of studio courses. Some schools assume high school mechanical drawing is a sufficient prerequisite. In general, quality control on student

[3] Here, architectural drawing is distinguished from sketching and doodling, which is a very informal and personal type of drawing, to be discussed later.

capabilities in drawing has been reduced in US architectural education. Yet studio courses, the center of an architectural curriculum, assume that students have adequate skills to use these representations in design work. Drawing competence is usually determined by the quality of the drawn product. Knowledge of how representations are used allows posing the competence question differently. Is learning to read drawings to the point of becoming an automatic process a necessary skill for effectively working in architecture? Is integration of multiple 2D drawings into a mental image of the 3D form a necessary skill for architectural expertise? Do expert architects rely being able to "read" drawings? We have all partially learned some external representations that we have not internalized, such as a foreign language or some form of higher mathematics. Such examples point out the varying degrees to which we learn external representations and our differing abilities to reason or think in them.

The special representations learned by engineers and architects are not exclusive. Architects also learn new symbolic representations, including calculus. It is not expected that architects will often encounter calculus-based representations in their professional work, however. Similarly, most engineering fields learn some form of orthographic drawing, usually using a CAD program. Whether engineering students are expected to be able to transparently read, and mentally interpret, integrate and manipulate shapes and compositions, represented in drawings, varies from school to school.

The point is not that interpreting calculus formulas or architectural drawings are themselves examples of expertise. Rather the issue is whether understanding models represented in integral or differential equations and being able to reason about those models requires the more basic ability to easily read and interpret them. Similarly, the question in architecture is whether certain advanced capabilities, such as the functional and cultural interpretation of buildings, are based upon the ability to automatically read architectural drawings.

3.1 Design capability as a hierarchy of representational and reasoning skills

After gaining capabilities in calculus, engineering curricula requires students to undertake work in statics, thermodynamics, circuit flows and so forth to address analysis of physical phenomena. Students spend time learning the mathematical representations of a design's different behaviors with regard to forces, resistances, flow, to which a design responds. They learn to both characterize particular behavioral issues in terms of mathematical models, and also how to solve the formulated problems. They learn the measurements that correspond to the parameters of the models and how to extract these from real situations. They also are expected to understand and respect the limit conditions of the model and any other built-in assumptions. The different kinds of analysis are initially taught singularly. Later, they may be combined in more advanced courses that address integrating multiple basic forms of analysis in application areas, such as thermal stress, analog-digital circuits, electro-mechanical systems and so forth.

Similarly, the reading and interpretation of architectural drawings becomes a basic, assumed skill early in architectural education. Drawing reading is quickly assumed to lead to the integration of multiple views to interpret 3D building shapes and spaces. Later work builds upon these integration capabilities to deal with reasoning about the spatial qualities, functionality or use of the spaces defined in the multiple drawings. Students without the ability to mentally image 3D spaces may resort to making such assessments by constructing external representations of them. Repeated construction of an external 3D representation probably helps students to learn to mentally image spaces. In this way, the tasks have a potentially self-correcting structure. Only after learning to "read" and effectively use such external representations can issues of decision trade-offs, of values, risk and other high level judgmental issues be productively applied.

The skills used to address advanced engineering issues appear to build upon a complex set of previously learned capabilities. The composition of these capabilities define a lattice of skills, upon which more elaborate forms of reasoning are based.

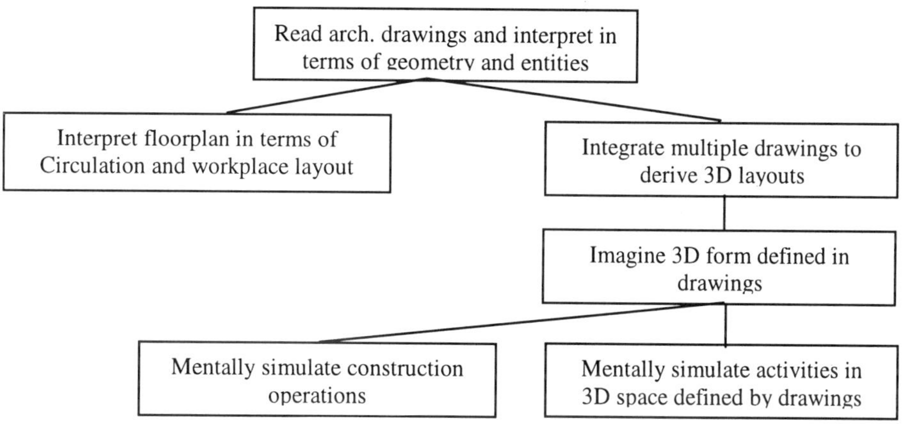

Figure 2: A partial semantic hierarchy identifying architectural drawing skills needed to mentally simulate portion of a design

Diagramming the individual capabilities required to carry out some high level design task allows us to better understand the atomic skills needed to accomplish the task. A small example is laid out in Figure 2. It identifies some of the prerequisite skills assumed in mental simulation of activities within a space, something expected of students in advanced architecture courses. (Other knowledge, for example regarding activities in the designed space, is omitted from the diagram.) The more basic skills are at the top of the figure, which are specialized by the capabilities below them.

Such a lattice articulates what needs to be known to possess particular design skills. Systematic study of the capabilities of students with regard to such hierarchies could lead to articulation and refinement of what the hierarchy of skills is and how we might facilitate learning its components. In addition, such articulation allows better diagnosis of student's problems. For example, why is a student not able to recognize visual or functional issues that arise within a specific layout? What is the cause of a student's generation of a building elevation based on a known style but that ignores basic proportional issues?

3.2 Use of mental imagery in solving design layout tasks

A large literature exists of anecdotal reports from people thought to be creative that report the contribution of mental imagery to their creative efforts. Some are by scientists: Einstein, Maxwell, Kekule, Watson, Bohr, Faraday, Pauli and Feynman, and others are by artists: Mozart, Wright, Surls (reported in (Intons-Peterson 93a; Miller 84; Shepard 78)). All of these examples deal with problems or issues that have an aspect that can be characterized spatially. Often, it is that spatial characterization that provides the crucial insight. Most areas of design, such as mechanical engineering and architecture, involve the spatial composition of entities. Spatial layouts are an important if not a central aspect in most areas of design.

At the cognitive level, spatial layouts may be generated using different processes. The range of processes may be defined by two extremes. At one extreme, the design may be fully generated in an externally drawn representation. It is composed and refined by manipulating the symbols and structure of the drawing. In this case, the designer's knowledge consists of procedural operations that manipulate the drawing and applies various criteria regarding an appropriate layout. A variation of the first extreme is using a master layout to record major decisions, but to generate small more detailed but partial layouts to deal with aspects of the design. That is, external representations are used, but they are hierarchically decomposed, then recomposed. At the other extreme, the designer can build up the design layout mentally in his or her head, applying manipulation operations and criteria mentally to the layout, until an appropriate layout is generated. The external representation is later used to record the layout generated mentally. In between these extremes are various mixed strategies where portions of the design are generated mentally, then added to the external drawing.

A famous anecdotal example of the second strategy was reported by Frank Lloyd Wright's staff at Taliesin. Mr Wright had been hired by Edgar Kaufmann of Pittsburgh to design a summer home at Bear Run, in central Pennsylvania. Mr. Kaufmann telephoned Mr. Wright from time to time to learn the progress of the project and Mr. Wright always reported that the project was proceeding well. No drawings were generated however. This type of communication continued for over a year, with Mr. Wright always reporting that the project was proceeding well. After about 15 months, Mr. Kaufmann called unexpectedly and asked the same questions. When Mr. Wright reported that the project was proceeding well (again), Mr. Kaufmann told him he was nearby and would like to come by and review the drawings. He said he would be there in an hour and a half. Mr. Wright said fine and hung up, then sat down and in about an hour drew up the plans, very near to their final form. He design became known as Fallingwater, one of the most famous residential buildings in America. He had clearly developed the design fully in his head, and then produced the drawings to record it for the client[4]. Wright wrote about his design process in 1928.

> "Conceive the building in the imagination, not on paper but the mind, thoroughly—before touching paper....Let it live there—gradually taking more definite form before committing it the draughting board. When the thing lives for you—start to plan it with tools. Not before. To draw during conception or sketch, as we say, experimenting with practical adjustments to scale is well enough if the conception is clear enough to be firmly held...But if the original concept is lost as the drawing proceeds, throw all away and begin afresh" (Wright 28).

The report on the design of Fallingwater seems to be an exceptional case of developing a full design in ones head. It and Wright's admonition and the other creative breakthroughs mentioned earlier, seem to argue that the development of designs mentally, with minimal reliance on external representations, is the desired mode of creative design. But is this true? Does creativity rely on mental imaging? Or are these examples really pointing out people who either have unusual imaging capabilities (Wright) or have had to resort to such activities out of necessity because they could not draw (some scientists)? Do better results usually occur with greater reliance on external representations? Or is creative ability significantly enhanced through mental imagery?

Finke (90) has carried out a number of studies that show that average people are able to use mental imagery to compose forms and derive emergent practical objects from them. He had them memorize shapes such as those shown in Figure 3. He then asked them to make interesting compositions in their heads of the memorized shapes, without reference to any function.

---

[4] This event was described in the Public Broadcasting Service special biography of F.L. Wright, televised in 1998. It is also reported in Tafel (79:p3-7).

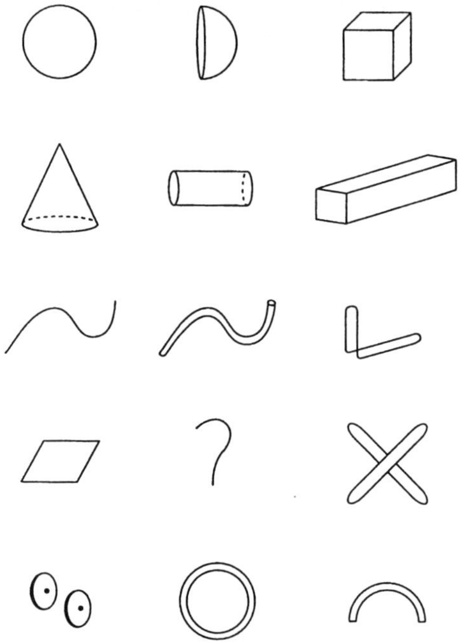

*Figure Three: Component objects and parts used in creative invention experiments.**

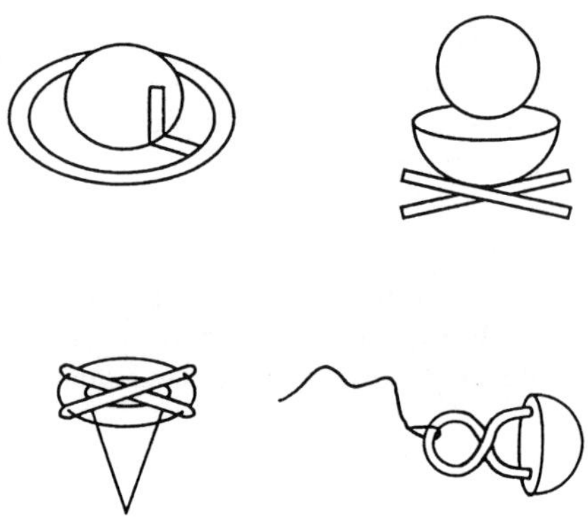

*Figure 4: Example compositions generated in the creative invention experiments.**

* Figures 3 and 4 are from Finke (90), with copyright permission.

Examples of the composed shapes are shown in Figure 4. He then gave them categories of uses—furniture, personal item, transportation—and asked subjects to apply the uses to the composed shapes. The number of objects identified as applicable were small, since they were created without the assigned function in mind. However, the objects defined were independently considered as very creative.

These studies have been used to argue that mental imaging can enhance creativity in average people, further reinforcing the mental imagery hypothesis. While studies have shown that mental imagery can be enhanced through training (Wallace and Hofelich 92), the effect of such training on creativity has not been tied together. Shepard (78) sought to show a direct linkage between creativity and the ability to mentally image but was only partially successful. Tests of children for correlations between creativity and IQ show a positive correlation, but these results are hard to separate from other correlations with IQ (Intons-Peterson 93a).

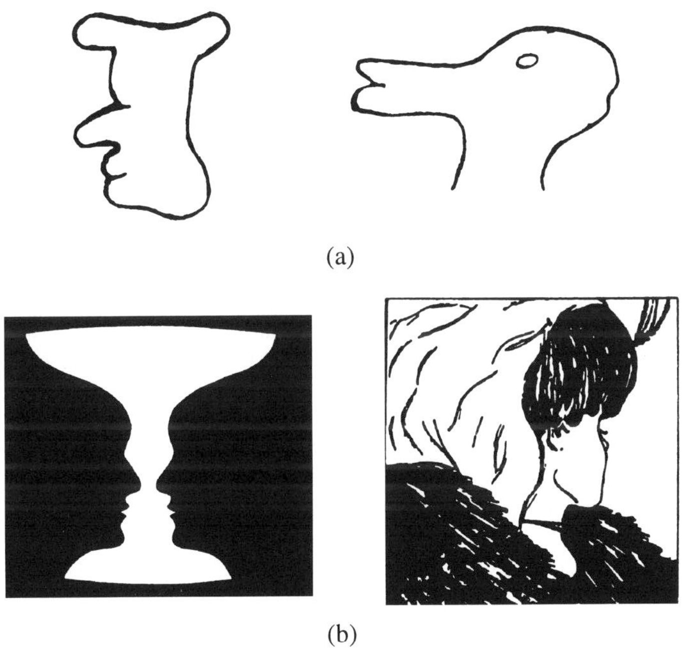

Figure 5: ambiguous figures used by to identify reference frames in imagery.

Other results that have been interpreted as showing that mental imagery can limit creativity. An often-cited study by Chambers and Reisberg (85) showed that mental imaging tends to fix the interpretation of the image. They had subjects memorize ambiguous figures, such as those shown in Figure 5, which were named to give a particular interpretation. Later when asked to imagine the shape and find another interpretation, they could not. However, if they were asked to draw the image, either on paper or virtually in their head, a significant number of subjects could re-interpret the image. These results were interpreted to show that the drawing could be reversed if it became a perceptual process. These results were construed to show that imaging relies on interpreted images, as distinct from less processed perceptions.

A number of studies have elaborated and further interpreted these results (Reisberg and Logie 93; Peterson 93). The current understanding is that mental images are interpreted within a *reference frame*. The reference frame includes orientation, scale, figure-ground relations, and classification of features (Chambers 93). Later operations on the mental image can be undertaken if they do not require changing the reference frame. The first form in Figure 5(a) can be interpreted as a chef looking to the left, or an upside down dog. The second form in Figure 5(a) can be interpreted as a duck looking left or a rabbit looking upper right. Both figures require change of the reference frame. In the left form in Figure 5(b), the martini glass or the two faces interpretations do not require changing the reference frame, nor does the young lady or old lady in the right image. It was found that the images in Figure 5(b) are easier to re-interpret mentally than those in Figure 5(a). Self-conscious action is required to alter an image's reference frame. Thus in certain situations, mental imaging may fixate thinking according to one interpretation.

From these and other studies, it seems that at least two kinds of actions can reduce such mental fixations. Asking a person to draw the figure seems to have a loosening effect, facilitating re-interpretation (Chambers 93). Also, if a person is instructed (by others or oneself) to re-interpret the mental image, re-interpretation may be undertaken. These results suggest ways to address the recognized problem of fixation in design, especially among beginning designers (Sachs 99; Atman and Chimka et al. 99).

These studies support Wright's contention. Mental imagery seems to facilitate design creation, especially in generating new concepts or compositions. Through imagery, people can learn new things from images they have generated. Mental imaging may be limiting, however, if relied on alone (without sketching) and if re-organization of the mental design concept is required (Reisberg and Logie 93).

However, most design protocols don't describe designers thinking about their design contemplatively. Rather they describe designers as active, frequently sketching, making notes, generating layout drawings, describing issues, context and forms. Donald Schon's book and papers present and analyze protocols of Quist, (Shon 83, c 4), then Petra and Clara (Schon and Wiggins 92; Schon 92) that capture the dynamics of such processes. The analyses by Schon are different from most in that their interpretation seems to correspond more closely with what designers themselves think they are doing. Schon makes several important observations about the process of design, drawn from the protocols. One of them is the dialogue that a designer carries out between their internal mental representation of the problem and their external representations. This dialogue involves "moves" that take conceptual positions in response to the mentally understood context (slope of site, population, climate, social organization and activities). Moves seem to apply to both the internal representation and the external ones. He interprets how a sequence of moves—what he describes as "seeing, moving, seeing" build upon the context of earlier moves. Later he also shows how a form has multiple interpretations and is conceptually decomposed in different ways, for example as multiple letter Ls or as a cluster with attached blocks. These different emergent structures lead to the remembering of different design issues and different moves. He also notes how a designer mentally simulates walking along a path to derive new design criteria.

Reading these dialogues carefully, one gets the impression that Schon's "seeing" involves internally representing spaces and forms that are drawn externally, then a "move" acts upon the two representations, either correcting undesired conditions or elaborating the design. "Seeing" again leads to repeating the cycle. Rather than the two extremes, this type of design process seems to rely heavily on a "dialogue" between mental representations and simple external representations that abstractly capture some geometric properties and relations, while most of the context is carried mentally. Sometimes a move is a change of form, other times it is a reframing or reinterpretation of the external representation, without a physical change.

Many design fields and designers revere sketching as central to creativity (Robbins 94; Henderson 99), but we know only a little about how sketches operate cognitively. A similar analysis of the tasks used in the Schon study[5], but focusing on the use of sketching, is presented is Goldschmidt (91). Goldschmidt identifies a "dialectics of sketching" that cycles between "seeing-as" and "seeing-that". She considers these actions from a perceptual and imaging standpoint. Seeing-as is a perceptual interpretation of a sketch, structuring it in various ways. Seeing-that applies non-figural arguments to the sketch. In other words, sketching generates new forms (or allows re-interpretation of an existing one) that are interpreted by the mental imaging process, which leads to revision of the sketch. Later Goldschmidt extends this viewpoint (94), presenting arguments that early phase "doodling" on paper is a form of interactive imagery. She proposes that doodling supports and augments mental imagery, allowing the quick formation and manipulation of complex forms, forms that would be challenging to generate or maintain mentally. The sketches provide cues for retrieving new information and for re-interpretation. Such doodling can be vague in the way that mental images have been shown to be vague (Chambers 93).

[5] The exercise used by Schon and Goldschmidt was originally developed by William Porter and is discussed at length in Schon (88).

Goldschmidt's hypothesis that doodling is an extension of mental imagery is provocative regarding the close link it proposes with imagery. If we are to understand sketching it needs to be tested. The hypothesis seems to assume that there is a one-to-one correspondence between mental imagery and sketching. Do designers mentally capture the image of sketches as they are created? As an example test, can they be distracted after generating a doodle and still reproduce it? Is a large doodle structured in a way similar to images, as a series of possibly overlapping images, as proposed by Kosslyn? The relation of sketching to mental imaging is especially interesting in light of the previously reviewed work indicating that images are necessarily interpreted while drawings need not be. How does interpretation apply to this dynamic process of mental and physical image generation? For some further discussion, see Goldschmidt's Chapter this volume[6].

Schon and Goldschmidt have begun to identify a mixed role for mental imaging, its interplay with symbolic or propositional knowledge, and the interaction between imaging and sketching. The processes of design described by Schon and Goldschmidt however are very different from those described by Wright. And these are different from the processes reported in most protocol studies. Are all these approaches equally effective? Further investigation of these questions may allow us to better understand design thinking and how to it can be more easily learned.

3.3 Design Tools as External Representations

The cognitive process by which people learn language and mathematical representations provides a useful perspective on how they *should* learn computer-based design tools. Computer-aided design (CAD) systems are becoming standard tools-of-the-trade for architects and the basic tool for geometric representation by mechanical and other engineers.

The negative side of current CAD systems is the user interface complexity of these tools. Architects and engineers have strongly criticized most CAD drawing and modeling tools because of their complexity and long learning time (Johnson 97, Henderson 99:c6). The complexity of most CAD systems is such that without daily use, constant attention to how to operate it is required. Users must spend mental effort at the user interface of the tool in order to carry out most operations, distracting them from design issues. An important goal of the field of human-computer interaction is to reduce the cognitive cost of computer interfaces so they can be used transparently, with their operations being automated by the user, as in writing and typing. Can this goal be widely achieved for CAD systems also? It would seem to be mandatory if computer-based sketching is to support the role posed by Goldschmidt (94) of augmenting mental imagery. She identifies some criteria for such

[6] For a survey of current work in sketching, see Purcell (98).

sketches: economical, fast, ambiguous and easily managed over time. More generally, all general purpose tools that are meant to support design need to be easily learned to the point of automatization. This is an important criterion that explains many of the limitations of current CAD tools.

There has been anecdotal reports that knowledge of form manipulation enhances the ability to generate and interpret form, what architects might call "form literacy". Sculpting and printmaking were basic parts of the Bauhaus curriculum, in support of these assumptions (Dearstyne 86). Do students that have learned well a solid modeling tool, such as Pro-Engineer®, 3D Studio Viz® or Architectural Desktop®, also become familiar with complex shapes, in the way a sculptor does? Do such students gain new ways to conceive of shapes in their heads? Do they learn to carry out sculpting operations in their heads, as might a sculptor? These questions apply to both architects and engineers. A possible benefit of these powerful sculpting tools is that they may allow users to become more knowledgeable in forming, recognizing and interpreting complex shapes. These hypotheses have not yet been tested.

3.4 Methodological Issues

A variety of approaches have been developed to study imagery and other mental representations. Most have relied on experimental methods. Some seem well suited to studies in design cognition. In an experimental setting, it is desirable to simplify responses of subjects to simple tests.

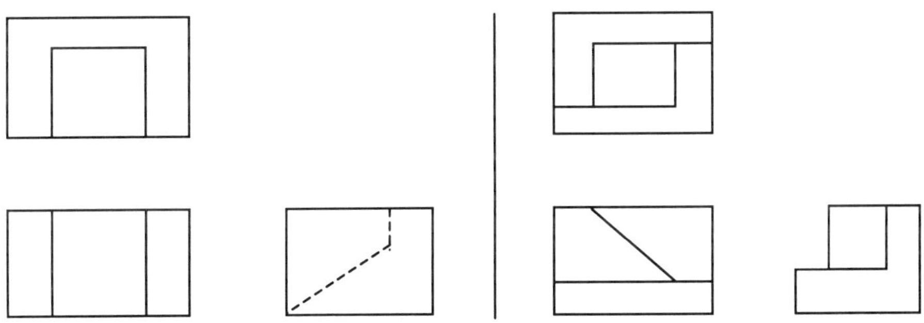

Figure 6: Multiple orthographic views of shapes. Are they consistent or do they describe an impossible shape?

One often-used procedure asks whether two or three different views of an object are of the same object or different ones. Two example objects are presented in Figure 6. The views may be given simultaneously or sequentially (Cooper 89). The assumption is that answers require the subject to mentally integrate the two views.

Another test is to present two or three orthographic views, then ask subjects to match their mental image with an isometric view (Cooper, Mumaw and Morrow 84) or carry out operations on the 3D shape. These tests require subjects to integrate multiple views, then to determine if they are consistent or to carry out other forms of reasoning on the integrated 3D shapes.

Such experiments must be structured carefully. Other strategies may be used by subjects to generate responses a 3D integration. Checking the number and location of vertical or horizontal edges or blocks and the order of changes allows differences to be identified, without mentally generating a 3D model. Such a strategy is easily applied to the forms in

Figure 6. That is, symbolic operations can be applied to some visual tasks, which defeats their intention and leads to wrong interpretations of the experimental result. It is critical that tests be pre-tested with visually educated users, to assess the different strategies that may be applied.

A reasonable hypothesis regarding mental imaging is that one effect of a design education is to develop student skills in mental imaging. That is, a professional architect or industrial designer should have greater imaging capabilities than a senior design student. A senior student designer should have greater imaging capabilities than a freshman. Similar but possibly less strong effects should occur in engineering education. We know that mental imaging capabilities can be improved with training (Wallace and Hofelich 92; Seel and Dörr 94). However, we currently have no validated measures of the ability to reason and make inferences in a mental representation. The ability to assess student capabilities would be enhanced if we could identify reliable measures of this mental skill. A commonly used test to assess mental visualization skills in people is the Minnesota Paper Form Board (Likert and Quasha 41), but questions have arisen regarding these and other tests of imaging, because they have not correlated well with individual measures of imaging ability. See Paivio (86) and Intons-Peterson (93a). Perhaps carefully constructed tests that require imaging ability are the best tests for now.

An example test figure is shown in Figure 7. To test if students are able to read drawings, the questions can be asked with the drawing present. More challenging is to present the drawing and ask students to study it until they can remember it, then take it away and ask the questions. I propose that they can only be answered by carrying the image of the drawing in ones head.

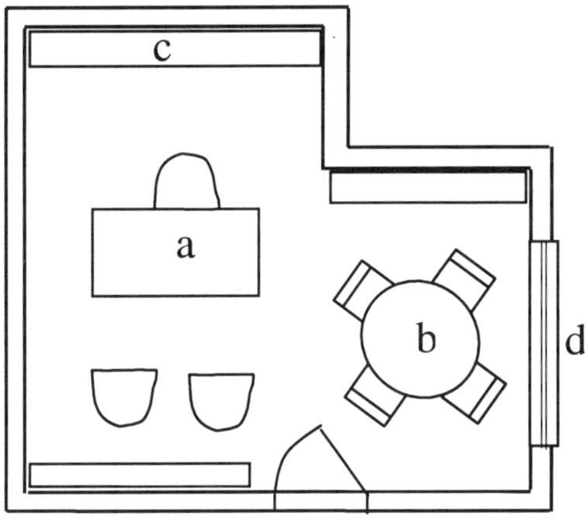

1. what are a, b, c, d in this drawing?
2. how many windows are there?
3. how many people can be seated?
4. the person at the desk can see the f
5. aces of how many people?
6. when the door is first opened, can a person at the table see who opens it?

Figure 7: An architectural drawing and some questions to be asked about

A more general question regarding mental representations is "at what point of learning any external representation does it become available for internal representation and support mental reasoning?" With repeated practice, "reading" an external representation becomes easier and eventually becomes transparent, allowing automatic interpretation. In roughly parallel fashion, practice allows us to gain the ability to mentally use the representation and to carry out simple mental operations. Do these two capabilities—automatic reading of a representation and the ability to mentally reason in the representation—come into being at the same time as complementary capabilities, or does one precede the other? It would be valuable to determine this relation, if it exists. If they took place together, then testing could use one capability as a surrogate for the other.

In summary, the development of expertise in the use of both internal and external representations is one of the hallmarks of design education. At the time of introduction to a new representation, people are very aware that they are learning something new. As their familiarity grows, using the representation becomes easier and its application more confident. Still later it becomes routinized; we might say it has become a skill. At some point, it becomes automatized and we are hardly aware of the representation and are able to focus on its semantic content. During this time, the ability to use the representation internally, in ones thinking, grows. As the representational skills advance, new skills are based on them. These may be based on reading the representation or manipulating it, and/or interpreting various results. It is likely to involve mapping between the representation and other knowledge, held for example verbally or mathematically. This seems to be the point where design expertise can emerge. The development of the lower level representational skills is a prerequisite for the high level reasoning and for the actions required of effective designers.

Hopefully, we can learn how such skills are built up and the critical links required in developing such a lattice of skills. Research in mental imagery in cognitive psychology allows us to frame many questions of relevance to various design fields. I propose that it offers an practical foundation for future work in design cognition.

Now I turn to review another important issue in design education, the structuring of a design context.

4 Structuring of Design Contexts and Recalling Relevant Design Features

An important aspect of any design task, recognized in both engineering and architecture, is the structuring of the design context. This can be characterized in different ways: for example, as Simon's ill-defined property, distinguishing design from other forms of problem solving; as Winograd's contextual "situatedness" or as the development of Schon's "design world". Described in more engineering terms, designers are assumed to be able to interpret a design problem readily into a set of performance or behavioral requirements of the technical system (to use Hubka and Eder's term (88)). That is, all descriptions of design characterize within their own vocabulary the important subtask of formulating an understanding of the context of the design situation, of the function and behaviors important to a successful design, to identify relations and dependencies, whether technical, user-defined, cultural, climatic or economic, and the class of materials, features, construction methods, or other entities the design is to be composed from.

Students in both engineering and architecture have some base knowledge of their field when they begin it, gained from living in buildings and visiting many others, or in engineering, gained from working on cars, computers or other machines. Previous learning, from personal reading or school, has also added to their mental "store".

Architectural education is assumed to incrementally build up a rich memory structure for young architecture students through studios, courses in architectural history and field trips, all considered essential to their education. Together these courses present a historical/cultural view of buildings, their function and response to context, as well as details of their form and aspects of their experiential qualities. The architectural curriculum is organized to structure the diversity of individual experience and to develop a shared way to attend to architectural experiences[7]. The encoding of architectural knowledge is a complex, multi-modal structure, mixing perceptual memories—visual, auditory, kinesthetic, possibly smell—combined with historical, critical assessment, and other propositional knowledge. In later studio projects, students are expected to be able to identify precedents and analogies with their work, as a way to explain and justify it.

A component of engineering design education is the design project, where students scope out and specify a product to design, in response to some given set of functions (often nonsense functions, such as catapulting an egg some distance and catching it without breaking). In this type of experience, students are expected to gain skills in abstractly defining a structure of functions and composing a mechanical system that supports them. If the project is done well, they will learn a vocabulary to identify different functions, various structures that respond to the functions and gain experience in composing structures. They will learn to identify the critical behavioral parameters, tune those parameters, and so forth. They may also learn to consider secondary functions, such as maintenance, safety, pollution that are important to successful designs in modern society (Ullman 92).

In both architectural and engineering design, the form vocabularies from which designs are composed are very large. For example in mechanics there are many types of mechanisms; in architecture, there are many building types, each with their own set of prototypes. It is not practical for students to become deeply familiar with all types while in school. Rather they generally gain experience in the meta-capability of how to effectively learn and use such taxonomies by examining one taxonomy or a small number. It is assumed that students will later know how to learn the detailed information about a particular design class when it is needed in practice. However, in both fields, consideration should be given as to whether the structuring of this knowledge is done effectively and the general skills appropriately conveyed. This applies to both the structuring of the data, taxonomically or otherwise, and its application and use.

Problem structuring capabilities are an important part of design education of all types. Protocol studies have shown that problem definitions generated by different designers are highly varied. My set of early room layout protocols varied by a 3:1 ratio in terms of relevant criteria generated (Eastman 69, 70). In this case, the criteria were based on the same explicitly given information, suggesting that differences in previous experience, the encoding of that experience and/or differential abilities to recall relevant experiences were influential in producing the results. The subject that generated more than twice as many

[7] One can question from this perspective the effectiveness of the heavy weight put on required history courses. Because the context of historical buildings is so different from current ones, many architecture students receive little analysis of current building types, such as shopping malls, government buildings, airports, high-rise offices, schools or hospitals.

criteria than the other three subjects tested offered an insight. After collecting the protocol, I asked him how he was able to identify so many criteria. Paraphrasing his answer, "Whenever I look at a design, I consider it from a basic set of issues. Bathrooms and stairs are the places in residences where most accidents occur. Also, cleaning of a bathroom is a major issue. A bathroom is difficult to keep clean and any reduction of cleaning problems should be attractive to purchasers. I apply these criteria as well as my experience whenever I review some design. That is, I look at new designs from an analytic point of view." This designer had a strategy for encoding design information whenever he looked at example designs. He used this richly encoded information when new design problems arose. This suggests that some designers have trained themselves to experience designs in a specialized way, so that they gain more useful design information from a given experience. This anecdotal description from a designer, who later became well-known industrial designer, suggests that design recall is a complex interaction between structuring experience, reflection about design knowledge and later recall in new design situations.

4.1 Recall of design information

Psychology provides an abstract interpretation of retrieval processes. The associations between information, regardless of modality, are thought to be the basis for recall. The term *activation* is used to identify the strength of associations. Only associations above a certain activation level enter consciousness. On the learning side, conscious attention to an association increases its activation level. However, associations that are experienced but not consciously attended to also increase the activation level, but to a lesser degree. Other influences on recall are the number and uniqueness of cues used to recall some piece of information, and the number of times the association has been used in recall (reviewed in Glass and Holyoak 83, c 7).

At the simplest level, naming a phenomenon or feature allows it to be recalled more easily (Barsalou 93). Architecture and the various branches of engineering have both developed specialized vocabularies to describe important concepts. Names typically denote an individual item or a class of similar items, providing associations with other similar concepts. It also has been hypothesized that imagery is a necessary component for classification, which allows naming (Intons-Peterson 93). That is, classification occurs through the matching of an image with the phenomena being classified. .If this is the case, then people who are good imagers, as architects are proposed to be, should also be good classifiers. This symmetry seems to support the idea that the name-image relation supports multiple means to recall information.

The rich cognitive structures into which design information is encoded, are assumed to facilitate recall. The benefits of such structures were verified by Mäntylä, who ran tests showing significant increases in recall as a result of developing multiple associations and distinctive access cues (Mäntylä 86). There is a well-developed body of research identifying various factors to improve recall (Kihlstrom 96). However, most emphasize encoding strategies and duplicating as far as possible the multiple cues existing during the encoding.

While we could go down a path of design cognition research that examines individual encoding and retrieval processes, it is not apparent that this path will lead to enhancement of understanding how people learn effective strategies for recalling information while designing. Recall implies finding something in memory that is already stored there. But structuring a design context is much more complex than recalling the right "fact" at the right time and seems to require more complex processes.

When an architect learns about a building, the kind of memory structure defined earlier gets recalled. Relations are defined with that knowledge to existing knowledge. If someone asks "what other buildings are like the New York Guggenheim Museum?", one looks for properties of the Guggenheim and searches for other buildings with those properties. If someone says that the Guggenheim is like a circular car park (not a very attractive association), this association is built from two different relations: a spiral circulation path and a cylindrical form. Mentally constructing such associations, finding common properties or processes in different phenomena, is called forming an *analogy*. Analogies have been an area of study in psychology for some time (Duncker 45; Gick and Holyoak 80; Gentner 89).

The classic analogy is described by Duncker (45) as the radiation problem—of how to destroy a tumor with radiation without destroying surrounding tissue. The source case tells of a general who instead of attacking a fortification straight on, disperses his soldiers all around the fort and has them attack from all angles. The attack succeeds because the soldiers in the fort did not see that an army was attacking them. From this example can you find how to solve the radiation problem? In this perspective, analogy is an example of memory retrieval, accessing indirectly related design information and making a new association. Once the association has been made, the analogy is no longer of interest; it is the making of the association that is the critical cognitive event.

Rowe cites the following examples of the use of analogy in architecture, drawn from the architectural literature: the roof of Le Corbusier's Ronchamp Chapel being shaped like a crabshell, the form of FL Wright's Unitarian Church shaped like hands folded in prayer, and Utzon's Sydney Opera House shaped like sails on a yacht (Rowe 87, p. 82). Engineers' use of functional analysis, that is decomposing a complex system into its individual functions, then using the functions to identify alternative mechanisms that might be used, is considered a form of analogical reasoning by Chi, Feltovich and Glaser (81). A number of studies of the use of analogies have been made in other fields (Vosniadou and Ortony 89).

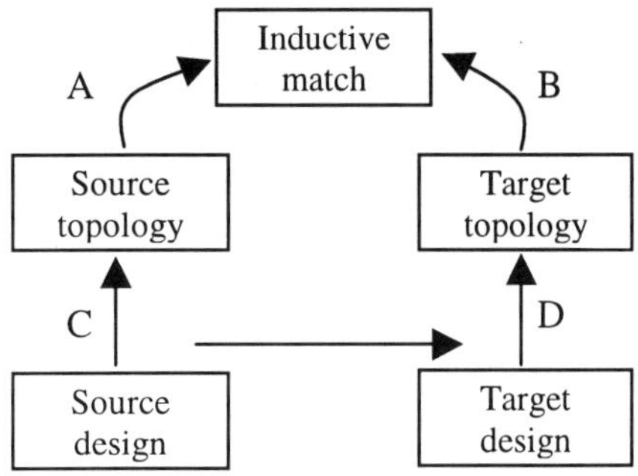

Figure 8: The structure of analogical reasoning.

An analogy arises when some relations in one situation can be adapted to address a problem in a quite different situation. An analogy is defined as a likeness relation, as in A:B as to C:D (shown in Figure 8). It involves an abstraction taken from one situation and applied to another. Analogies are different from similes and metaphors: similes are of the form "A is like B", where A and B have some properties in common; metaphors are an unconventional way of describing one thing in reference to another, based on some common semantics. Analogies involve relations that have some correspondence in two situations or things, called the *source* and the *target*. The source is the information from which the relation is extracted and the target is the information to which it is applied. Analogy is considered to consist of the following processes: (1) abstracting some significant relations of the target, (2) recalling or abstracting the relations of candidate sources, (3) making an inductive association between the target relations and those of some source (the A::B mapping), (4) identifying other relevant source properties or relations that can be mapped from the source to the target in a similar way that resolves a problem or unknown (the C::D mapping) (Vosniadou 89).

Analogies can be made by children as well as by adults (Holyoak, Jun and Billman 84). What is different are the relations by which analogies are identified. Analogical relations are described as being "shallow" or "deep". Shallow analogies are those made based on the perceptual or surface features of the two phenomena, such as their color or shape. Deep analogies are based on abstractions of the phenomena, such as their topological structure or function. Some refer to these correspondingly as iconic and structural analogies. The architectural analogies described by Rowe, reported earlier for Le Corbusier, Wright and Utzon, are all iconic analogies. Some people do not consider associations without relevant abstractions to be analogies, but similes.

Engineering involves the application of complex analogies. A high level goal is to identify and apply an appropriate math model characterizing a design's behavior. Errors arise if the wrong math model is applied, or its parameters used incorrectly. The application of an appropriate mathematical model to a design or aspect of a design is a multi-step process. The process has different contexts, requiring different actions. The pure synthesis process is the easier. A mechanism is selected and its associated mathematical model is applied; performance requirements are defined and design parameters selected or derived to meet them. In other cases, however, the design has already been generated, possibly in response to other behaviors or criteria, and the design generated must be analyzed from a particular behavioral perspective. This kind of real-world application of engineering concepts has been cited as a common weakness of student engineers.

This process is another version of the analogy process described above (Chi, Feltovich and Glaser 81). The association of a math model to a design involves two abstractions: of the existing design and of the candidate source mechanisms. The relation between the math model and the target abstraction is part of the initial learning. For lack of a known terminology, I call the source and target abstraction the source and target *topologies*. The topologies are needed to match the design with an appropriate math model. Once matched the corresponding math model can be applied.

Different representations of the topology of a model can be used in learning mechanisms and their math models: in words defining relations and properties or graphically as diagrams. A third alternative is that the second abstraction is not needed and that students can identify the appropriate mapping from the math model. Are there differences in the ability of students to apply mathematical models of behavior, based on the topology representations used in teaching the mechanisms? Studies of primary school students show that pictorial information is a more effective medium for presenting analogies than words or kinesthetic actions (Gineste 94). In engineering, are diagrams or word descriptions more effective than the math models alone, for students learning and application of what they know?

In previous decades, most engineering curricula included lab courses, where empirical studies were undertaken, verifying the relations between the behavior expressed in the mathematical model and the measured behavior of physical objects or mechanisms. By measuring the object, students directly experienced the relation between the model parameters and the physical object. They also learned the correspondence between the physical object and the mechanism topology. This part of engineering has been dropped from many curricula. It may have had unseen value, showing clearly the relation between mechanisms and associated parameters of a math model.

Analogy is an important form of recall process widely used in design. There have been only a few studies of its use in design and its range of use has not been well documented. Architects seem to become more adept at applying analogy as they become more expert (Casakin and Goldschmidt 99). Is this a tacit result of their design education? Or is it a filtering that eliminates aspiring designers that cannot form analogies? Can analogy use be enhanced by specific training? Some information may support the forming of analogies better than other information. What topological abstractions are most useful when the intention is to identify the correct behavioral model using analogy? Analysis of a specific mechanism or building from many perspectives and behaviors demonstrates how a mechanism has many associated abstractions. Such studies of an object are sometimes used in design. They would seem to be one way to show students how to abstract a design to generate more possibilities of analogical transfer.

4.2 Studies in Recall and the Use of Analogy

Comparing and studying designer's methods for recalling relevant design information can be studied using protocol analysis (Eastman 69; Dwarakanath and Blessing 96). In these cases, the original encoded information in memory is not known, but protocol studies can identify differences in recall and highlight interesting cases. But in order to gain a deeper understanding, more targeted studies seem required.

Most uses of protocol analysis in design have been to study the generation of a design from scratch. However, more constrained and structured forms of design study can be fruitful. Crismond (97) used an adaptation of a protocol study that elicits rich insight into how information is recalled and used. His adaptation involves a task structure called "investigate-and-redesign". It was developed to be a less time-consuming and quicker means to assess design skills, than the more common design-and-redesign tasks. It provides a rich means to assess what information is available and how it is used. The investigate-and-redesign task structure consists of six sequential activities:

(1) The subject investigates a set of existing products, all with the same function, handling and studying them, what Hawkins (74) referred to as "messing-about" The subjects assessed the product's function and identified its features.
(2) Subjects were asked to rank each product, using their own criteria, assessed by studying the tools directly, but not using them.
(3) They were then asked to use each product and to reflect upon what they had learned. Subjects ranked the items again, so as to note any changes in the rankings.
(4) They were asked to list features of an ideal device.
(5) Subjects were asked to design an experiment or user study, that would allow assessment of one or more products, as a way to learn what an ideal product might be.
(6) Last, Subjects were asked to design an improved product. This could be an incremental improvement of one of the products or a completely new product.

The two sets of tools to be redesigned—three nutcrackers and three jar openers—relied of physics principles for their functioning. The individual design samples were selected because of their novelty and potential unfamiliarity to the subjects. They included technical (mostly mechanical), aesthetic and cultural considerations in their design.

The subjects used in this study consisted of three groups. One group was of naïve designers from special high schools or colleges with special curricula in design. These subjects were selected for their reputation for being good designers. The second group, called "novice designers," were senior design students at well known universities, all of whom were recommended by their professors as being excellent designers. The last group involved experts. They were all experienced professional designers, inventors or design teachers. Most of this last group held patents and had multiple commercial designs to their credit. All studies were videotaped with two subjects working together, to facilitate verbalization. Each session covered the six activities and lasted two hours. Twenty-five pairs of subjects participated.

With existing designs as context, investigate-and-redesign scaffolds a space of designs and uses and potentially users, offering a motivating and grounded exercise. Among the insights about information use that Crismond found were:

- experts used analogies more than novices and naïve designers; they used deep analogies between the given object and other objects to determine the object's function ("looks like a clamp", "this is wrench-like", "just the same as a car-jack");
- each device involved principles of physics that partially determined how well they functioned (mechanical advantage gained from levers or screws); few naïve or beginning designers recognized these principles, while the experts recognized and applied them immediately;
- features of the product were poorly related to their function by naïve designers, while done richly, both qualitatively and quantitatively, by expert designers;
- all three groups discovered new secondary functions of the products during their use that were not identified by looking at it (some nutcrackers propensity to scatter nut shells rather than hold them: p. 90-91), the need to grip the jar as well as the lid in an opener (p. 252-3);
- naïve designers identified important features only while using it, while expert designers identified features during first examination of the product and used the features to infer the product's intended use;
- naïve designers identified design principles that were primarily derived from a user-perspective ("easy to use") while novice and expert designers identified criteria that were principles based ("I want to multiply the force");
- expert designers considered different user groups, such as arthritic users versus a professional chef (p. 232-3); naïve designer's primary user were themselves;
- in redesign, naïve designers predominantly work from an existing design and try to improve it; they also directly transferred aspects of an existing good design to reapply; those with more experience attempted to integrate abstract principles from multiple designs (p. 118, 127);

- making connections—between form and function, between high-level values and their low-level embedding in a product, between a mathematical model or physical principle and product features, application of an analogy to a product—were common in expert designers and rare or missing in naïve designers (p. 276-7);
- drawings were generally used as a scaffold to represent a device, used by experts and only a few novice designers; the drawings were used to support a dialogue with the design, as described by Schon and Wiggins (92), that extended their thinking about it (p. 278);
- for all groups, there were instances of ideas being recalled during one aspect of design, but forgotten and not applied later (p. 119-120).

Crismond's task structure elicited clear information showing how expert designers studied products using a variety of methods: taking them apart and putting them back together, testing cause-and-effect relationships, estimating forces and then testing them, studying how the product was assembled and predicting failure points. These allowed them to identify and recall a range of subtle criteria for the design of the household objects. Some expert designer's emphasized situated technology concepts, such as adjustment, discrete versus continuous, force concentration, portability, simplicity, while others relied on more science-based concepts, such as mass-force-acceleration, mechanical advantage, and distance-velocity-time relations. The experts used both vocabularies for describing products, their function and assembly, allowing them to identify design issues not recognized by the less experienced designers. The protocols articulated the expert's emphasis on keeping the design space open and not fixating early on any particular class of solution and to ask different questions to gain insight into the space of possible designs (Crismond 97, p 255-7).

Crismond's investigate-and-redesign task structure elicited information that allows better understanding how subjects recall and use design information. Asking about the designs initially, before their function is known, accessed the subject's knowledge about the artifact world, about forms and their relation with functions. Later, when the subjects used the products, criteria changed as a result of their experience. Asking the subjects to define the features and criteria for an ideal product of the given type, elicited information about how the subject has mentally linked functions and criteria with values. Last, re-design encouraged subjects to recall from their own repertoire of mechanisms and to study their integration capabilities. The investigate-and-redesign task structure allowed a rich range of information to be elicited from subjects with three levels of expertise. It should be applied to a range of designs, studying recall and use of information in a variety design domains.

Another study providing insight into recall was undertaken by Hernan Casakin (97). His Ph.D. thesis focused on the application of visual analogies in design. He was interested in whether they played a role in all types of design problems, or more so in ill-defined problems then well-defined problems. He was also interested in the differences in the use of analogies by beginning designers in comparison to experts.

Casakin's experiments consisted of timed design exercises (10-15 minutes), focusing on the development of one of several different design concepts. Five design tasks were used: two well-defined and three ill-defined. The well-defined design tasks were a staircase and a parking garage; the ill-defined design tasks were a viewing terrace, a prison, a dwelling. Here, ill-defined indicates a design task where the general form of the solution is open, while a well-defined task is one where the general form of the solution is fixed. All design tasks had an associated context and specification. Three groups of subjects were tested: beginning architecture students, advanced students and practicing architects.

Three different experiments were given.

(1) a design task with visual displays and explicit instructions to use analogy
(2) a design task with visual displays but *no* explicit instruction to use analogy
(3) a design task with no visual displays and no instructions to use analogy (control condition)

Each subject was given two design tasks, one well-defined and one ill-defined. Between 35 and 52 subjects participated in each of the three experiments, divided fairly evenly among the three experience levels. When visual displays were provided, these consisted of seventeen to twenty photographs, diagrams and sketches laid out roughly in a grid on a 1 x 0.7 meter board. The diagrams were both examples within the task design domain, as well as from other domains, as well as from science and art. They were selected separately for each of the five different design tasks to be loosely relevant, plus a few random fillers. The instruction to use analogy was given verbally, asking subjects to consider the displays and to use them in forming analogies to derive a solution. The design sessions were video taped and subjects were asked to think aloud.

Three independent judges at a foreign university made the evaluations and their scores averaged. The well-defined design tasks were scored on a binary scale, 0 for an incorrect solution and 1 for a correct one. The ill-defined tasks were scored on a 1 to 5 scale, separately for both design idea and for the solution. 1 or 2 meant that the design requirements were not met, and 3 to 5 were used to grade satisfactory designs. The reliability of judges' scores were assessed and found to be highly consistent.

For the well-defined tasks, the overall results were as one would expect; the beginning students had very low-levels of correct designs (between 7 and 54%), while the advanced students had generally higher levels (between 17 and 57%); professional architects had the most correct designs (between 18 and 83%). The figures are the range over the three different experiments, with the highest scores achieved when displays are provided and the instruction given to use analogies. The lowest scores were where neither were available —

the control condition. For the well-defined tasks, the visual displays significantly helped all three groups; the addition of the analogy instruction further improved the number of correct design results. The benefit resulting from the instruction to use analogy, however, was small for the professional architects and not statistically significant. This can be interpreted to mean either that their experience and the availability of the visual displays led them to use analogy without any instruction to do so, or alternatively that routine processes dominated their activities, resulting in them ignoring the use of analogy instruction. The benefits of the visual displays alone were also not statistically significant for the beginning or advanced students. That is, they seemed not yet ready to adapt or apply the visual materials to their designs, even though some of the visual materials were from the same design domain. The instruction to use analogy was necessary for the beginning students to make significant improvements.

TOTAL DESIGN		Beginning students	Advanced students	Architects
Analogy required Displays provided	Mean	3.463	3.734	3.984
	SD	1.036	.838	.537
No analogy required Displays provided	Mean	2.621	2.939	3.236
	SD	.758	1.029	.740
No analogy required Displays not provided	Mean	2.227	2.580	2.809
	SD	.689	.797	.822

Figure 9: Means and standard deviations from the three experiments for the three subject groups, scored for their total designs (1 to 5 scale) from (Casakin 97).

The ill-defined design tasks were scored on a 1-to-5 scale, requiring a different form of analysis from the well-defined tests. Total design scores are shown in Figure 9. Using unpaired T-Tests, the results, evaluating whether the effects of an instruction to use analogy versus having visual display but no instruction, were significant for all user groups, with <.002 or less probability. However, the results are surprising for the professional architects, since there were not parallel to the results for well-defined problems. The effective use of the visual information by practicing architects was not apparent without explicit instruction, for ill-defined problems. For the evaluation whether visual displays alone (without instruction) helped designers versus having no displays or instructions, the results for all user groups were positive, with <.092 or less p. That is, the evaluation scores for each group improved significantly when there was access to the visual displays. A third evaluation was made whether level of expertise was significant in whether designers were able to take advantage of the visual displays. The results were again positive, with significant increases in evaluation scores from beginners to advanced students, and also from advanced students to practitioners.

Additional tests were made to assess the relation between level of expertise and the use of analogy and visual displays. For students, it was found that the availability of displays with the analogy instruction provides a significant benefit for both tasks, but larger for ill-defined problems than well-defined ones. This result may be partially explained because it seemed that beginning architects treated the well-defined problems as being as difficult as the ill-defined ones; they did not have the routines available for solving these structured tasks. Also, beginning students seemed to not take advantage of the potential help offered by visual displays alone and only used them when instructed to do so—which led to significantly improved designs. This suggests that there is an educational opportunity to improve student design capabilities through the use of visual analogical displays. The benefits gained by professional architects were positive, but small and statistically insignificant when comparing their results for well-defined and ill-defined design tasks. This indicates that experienced architects solve both problem types the same way; both may be defined using analogies, or a familiar method may be utilized for solving both types of tasks, ignoring the instruction. Which of these two cases is the correct one needs to be determined [8]. These results are also reviewed and interpreted in Goldschmidt's chapter in this volume.

The task structure used by Casakin was different from most studies of analogy. Rather than studying the learning of specific analogies, these tests consisted of testing the more general issue of whether applying analogies was beneficial for solving a design task. The results support the general premise that analogies benefit problem solving, but leaves for future refinement the details regarding what operations designers use in carrying out the forming of analogies (but see Gineste 94; Vosniadou 89).

In the given experiments, the visual materials were available during the design time; recall from long term memory was not required for the source. A time lapse between seeing the visual materials and doing the design should indicate a decay rate in the benefit of the visual materials over time. How would the results of the design evaluations vary if the materials were studied earlier, a few hours, a day, or a week earlier? By putting a gap between the study period and the design task, encoding in long term memory and recall strategies come into play. This would allow understanding of the use of visual analogy in various different design situations. It may also allow us to learn better how design information may be abstracted when it is experienced to later facilitate analogies.

The analogies studied by Casakin were presented visually, suggesting the visual images were all processed iconically, as "shallow" analogies. However, some displays included compositional information and may have been used in a more structural or "deep" way. An extension of Casakin's work would be to determine whether successful applications of

[8] Another partial review of Casakin's thesis is given in Casakin and Goldschmidt (99).

analogy relied on structural relations. Possibly the failure of beginning students was because they did not know how to generate such analogies and attempted to rely on iconic ones.

As a last example study, I consider one from area outside of design, but that has implication for design expertise. In a study of medical expertise carried out by Feltovich, Johnson et al (84), they presented charts of cardiology cases to two groups, fourth year medical students and cardiologists with at least twenty years experience. The subjects studied the cases and reported aloud their thoughts about the case. They then compared the two sets of subjects. Medical students proposed diagnoses on a few symptoms, sometimes combining symptoms with inappropriate diseases (factual errors) and in other cases weighting symptoms either too heavily or lightly (judgmental errors). The experts did not associate symptoms with specific diseases, but with classes of disease. They evaluated symptoms relative to a large set of plausible diseases and explicitly considered potential alternatives. From the results of the study, Feltovich and his colleagues construed that medical students are familiar with a few limited prototypical cases of a disease, not the range of possible variations. Expert memories not only included knowledge of more diseases, but how to assess alternatives.

In medicine, symptoms are abstractions of individual cases and much of the work of making analogies has been undertaken communally within the domain. Thus the study dealt more with the content of the doctor's knowledge.

In some ways, the diagnosis of a disease from symptoms is similar to selecting a mechanism for a set of functions, or selecting an appropriate typological class for a building design. Designer should be able to recall various mechanisms not as individual solutions but as classes that respond to the given functions. They also should be able to distinguish between various typologies of buildings or mechanisms, based on detailed criteria, such as site conditions, cost, construction time, spatial differences for the building and based on such criteria as costs, fabrication, assembly and maintenance for the mechanism selection. Expertise is involved in how various typologies are evaluated and how subclasses are discriminated. This suggests another line of research exploring how designers use their knowledge in design contexts that could potentially generate valuable results.

The two main studies reported here both allow for fresh ways of looking at memory and recall in design. They control for previous knowledge in different ways. The Crismond study (97) sought to explicate the previous knowledge used by designers with different experience levels. Casakin (97) provided visual materials during the design task, but had to deal with the uncertainty that experienced architects already possessed the information provided (the displays and/or the instructions).

How does the relevant context and criteria for a design get identified and applied? The studies here begin to document in a rich way what is retrieved, and through extensive information about general knowledge, suggest how expert designers carry out these tasks in relation to beginning ones. While probing the range and extent of a designer's knowledge is difficult, these studies begin to cast a light onto their contents and organization.

5 Summary

The purpose of this chapter has been to identify some new issues of importance for design education and to point to psychology literature that suggests how these issues may be studied. I have attempted to delve deeply into two issues of interest. One is the use of mental imagery in the development of a design and its relation to external representations. I have assumed that both effective use of external representations also and learning them to the point that they can be effectively used internally as mental representations are basic skills that support more advanced work in all design fields. By understanding how this learning takes place and how different representations interact, we may come to understand the process by which these representation-dependent capabilities are built up, both within school and during professional experience.

The other area I have investigated is the interplay between personal experience and recall of design-relevant information. The experience may have been in the guise of education or more personal first-hand experience. Recall in design pertains not so much to factual recall as to responding to contextual conditions, accessed through the use of analogy and other complex associations. There is good evidence that outstanding designers both process their experiences differently from non-designers, and then use their experience in unique and powerful ways. Better understanding of their use of such experiential information resources may help in the education of designers that are more competent and creative, and better able to build upon and utilize their experience.

This chapter has made an initial attempt to integrate recent results in cognitive psychology with studies of design cognition and to pose both questions and identify research methods that may allow issues of design learning to be further illuminated. The survey presented has been motivated by the desire to move beyond understanding *what* designers do, as largely studied in protocol analyses over the last thirty years, to understanding *how* designers achieve their results. Of course there is no one answer to the question "how", in the same way that there may not be any one design solution to a given design context. Rather we are still interested in the range of effective processes designers use, how those processes operate and the skills that expert designers bring to bear in their work.

The approach to studying design and the kinds of knowledge it seeks is a continuation of the cognitive psychology tradition of information processing theories of thinking. It however recognizes the huge layer of knowledge based on previous cultural, social, interpersonal as well as formal knowledge about the world and how this knowledge is heavily drawn upon in reasoning during design. I have attempted to broaden the range of information and experience that can be included in the study of design cognition, in response to the criticism that design cognition has not addressed the situational richness that embodies design expertise. It will hopefully be possible to broaden our objective knowledge about design thinking and at the same time address the individualistic, cultural and social aspects that have been raised as counter-positions to the previous work.

NOTE: This paper has benefited from helpful comments from Gabriela Goldschmidt, David Craig, Fehmi Dogan and David Crismond. They made it better. The remaining limitations are those of the author.

References

Akin Ö, (1978), How do architects design? Latombe J-C, Ed. *Artificial Intelligence and Pattern Recognition in Computer-Aided Design*, North Holland Press, New York, NY, 65-104.

Akin Ö, (1986), *Psychology of Architectural Design*, Pion Press, London.

Alexander C, (1964), *Notes on the Synthesis of Form*, Harvard University Press, Cambridge, MA.

Atman C, Chimka J, Bursic K, and Nachtmann H, (1999), A comparison of freshman and senior engineering design processes, *Design Studies,* 20(3), (March) 131-152.

Ballesteros S, (1994), Cognitive Approaches to Human Perception: introduction, Ballesteros S, Ed. *Cognitive Approaches to Human Perception*, Lawrence Erlbaum and Associates, Hillsdale, NJ, 1-20.

Ballay J, (1987), An experimental view of the design process, Rouse W, and Boff K, Eds. *System Design: Behavioral perspectives on designers, tools and organizations*, North-Hollans Press, NY, 65-82.

Barsalou L, (1993), Flexibility, structure, and linguistic vagary in concepts: manifestation of a compositional system of perceptual symbols, Collins A, Gathercole S, Conway M, Morris P, Eds. *Theories of Memory*, Lawrance Erlbaum and Associates, Hillsdale, NJ, 29-102.

Baya V, and Leifer L, (1996), Understanding Information Management in Conceptual Design, Cross N, Christiaans H, Dorst K, Eds. *Analyzing Design Activity*, John Wiley & Sons, Chichester, 151-168.

Baykan C, (1996), Design Strategies, Cross N, Christiaans H, Dorst K, Eds. *Analyzing Design Activity*, John Wiley & Sons, 1996 Chichester, 133-150.

Bregman A, (1995), Demonstrations of auditory scene analysis: the perceptual organization of sound, Auditory Perception laboratory, McGill University, Montreal, PQ (digital disc and manual).

Brereton M, Cannon D, Mabogunje A, and Liefer L, (1996), Collaboration in design teams: How social interaction shapes a product, Cross N, Christiaans H, Dorst K, Eds. *Analyzing Design Activity*, John Wiley & Sons, 1996, Chichester, 319-342.

Candy L, and Edmonds E, (1996), Creative design of the Lotus bicycle, *Design Studies*, 17(1), 71-90.

Casakin H, (1997), *The Role of Analogy and Visual Displays in Architectural Design*, Ph.D. thesis, Technion – Isreal Institute of Technology, September.

Casakin H, and Goldschmidt G, (1999), Expertise and the use of visual analogy: implications for design education, *Design Studies* 20(2), (March) 153-175.

Chambers D, (1993), Images are both depictive and descriptive, Roskos-Ewoldsen B, Intons-Peterson M, and Anderson R, Eds. *Imagery, Creativity and Discovery: A cognitive perspective*, North-Holland Press, Amsterdam, 77-98.

Chambers D, and Reisberg D, (1985), Can mental images be ambiguous? *Journal of Experimental Psychology: Human perception and performance*, 11, 317-328.

Chi MTH, Feltovich P, and Glaser R, (1981), Categorization and representation of physics problems by experts and novices, *Cognitive Science,* 5, 121-152.

Chi MTH, (1997), Quantifying Qualitative Analyses of Verbal Data: A Practical Guide, *The Journal of the Learning Sciences*, 6(3), 271-315, Lawrence Erlbaum and Associates.

Craik F, and Lockhart R, (1972), Levels of processing: A framework for memory research, *Journal of Verbal Learning and Verbal Behavior*, 11, 671-684.

Crismond D, (1997), *Investigate-and-Redesign as a Context for Learning and Doing Science and Technology: a study of naïve, novice and expert high school and adult designers doing product comparisons and redesign tasks*, thesis, Harvard Graduate School of Education, Cambridge, MA.

Cooper L, and Shepard R, (1973), Chronometric studies of the rotation of mental images, Chase W, Ed. *Visual Information Processing*, Academic Press, NY, 75-176.

Cooper L, (1989). Mental models of the structure of visual objects, Shepp B, and Ballesteros S, *Object Perception: structure and process*, Lawrence Erlbaum and Associates, NJ, 91-119.

Cross N, and Cross A, (1996), Observations of teamwork and social processes in design, Cross N, Christiaans H, Dorst K, Eds. *Analyzing Design Activity*, John Wiley & Sons, Chichester, 291-318.

Crowder R, (1993), Systems and principles in memory theory: another critique of pure memory, Collins A, Gathercole S, Conway M, Morris P, Eds. *Theories of Memory*, Lawrance Erlbaum and Associates, Hillsdale, 139-162.

Davies S, (1995), Effects of Concurrent Verbalization on Design Problem Solving, *Design Studies*, 102-116.

Dearstyne H, (1986), *Inside the Bauhaus*, Rizzoli, NY.

Denis M, (1996), Imagery and the Description of Spatial Configurations, de Vega M, Intons-Peterson M, Johnson-Laird P, Denis M, Marchark M, Eds. *Models of Visuospatial Cognition*, Oxford University Press, 1996, 128-197.

Downing F, (1987), Imagery and the structure of design inquiry, *Journal of Mental Imagery*, 11(1), 61-86.

de Vega M, Marschark M, Intons-Peterson M, Johnson-Laird P and Denis M, (1996). Represent-ations of Visuospatial Cognition: A Discussion, de Vega M, Intons-Peterson M, Johnson-Laird P, Denis M, Eds. *Models of Visuospatial Cognition*, Oxford University Press, 1996, 198-226.

Dorst K, (1997), *Describing Design: a comparison of paradigms*, Ph.D. thesis, TU Delft, 1997.

Duncker K, (1945), On problem solving, *Psychological Monographs*, 58, 270.

Dwarakanath S, and Blessing L, (1996), Ingredient of the Design Process: a comparison between group and individual work, in Cross N, Christiaans H, Dorst K, Eds. *Analyzing Design Activity*, John Wiley & Sons, Chichester, 93-116.

Eastman C, (1969), Cognitive processes and ill-defined problems: a case study from design, Walker D, and Norton L, Eds. *proceedings Joint International Conference on Artificial Intelligence*, ACM, Washington DC.

Eastman C, (1970), On the analysis of intuitive design processes, Moore G, Ed. *Emerging Methods in Environmental Design and Planning*, MIT Press, Cambridge, MA. 21-37.

Emkin L, (1995), The threat of computers to engineering education, invited lecture, *Computers in Civil Engineering Conference*, Atlanta, GA.

Ericsson K, Simon H, (1984), *Protocol Analysis: Verbal reports as data*, MIT Press, Cambridge MA.

Ericsson K, and Charness N, (1994), Expert performance: Its structure and acquisition, *American Psychologist*, 49, 725-747.

Farah M, Soso M, Dasheiff R, (1992), Visual angle of the mind's eye before and after unilateral occipital lobectomy, *Journal of Experimental Psychology: Human Perception and Performance*, 18, 241-246.

Feltovich P, Johnson P, Moller J, Swanson D, (1984), LCS: The role and development of medical knowledge in diagnostic expertise, in Clancey W, and Shortliffe E, Eds. *Readings in Medical Artificial Intelligence*, Addison-Wesley, Reading, MA, 275-319.

Finke R, (1990), *Creative Imagery: discover and, invention in visualization*, Lawrence Erlbaum and Associates, Hillsdale, NJ.

Fodor J, (1975), *The Language of Thought*, Crowell, NY.

Foos P, (1980), Constructing cognitive maps from sentences, *Journal of Experimental Psychology: Human Learning and Memory*, 6, 51-78.

Fox P, Mintun M, Raichle M, Meizin F, Allman J, and Van Essen D, (1986), Mapping human visual cortex with positron emission tomography, *Nature,* 323, 806-809.

Feigenbaum E, and Feldman J, Eds. (1963), *Computers and Thought*, McGraw-Hill, NY.

Gentner D, (1989), The mechanisms of analogical learning, Vosniadou S, and Ortnony A, Eds. *Similarity and Analogical Reasoning*, Cambridge University Press, Cambridge, MA, 199-239.

Gick M, and Holyoak K, (1980), Analogical problem solving, *Cognitive Psychology*, 12, 306-355.

Gineste M, (1994), visualized analogies and memory for new information in first graders, Schnotz W and Kulhavy R, Eds. *Comprehension in Graphics*, North-Holland Press, Amsterdam, 251-268.

Glass A, and Holyoak K, (1986), *Cognition*, 2nd Ed. Random House, New York, NY.

Goel A, Gomez A, Grue N, Murdock W, Recker M, and Govindaraj T, (1996), Design Explanations in Interactive Design Environments, Gero J, Ed. *Proc. Fourth International Conference on AI in Design*, Palo Alto, CA, June. 1996.

Goel, V, (1995), *Sketches of Thought*, MIT Press, Cambridge, MA.

Goldschmidt G, (1991), The dialectics of sketching, *Creativity Research Journal*, 4(2), 123-143.

Goldschmidt G, (1994), On Visual Design Thinking: the vis kids of architecture, *Design Studies*, 15(2), 158-174.

Greenwood D, (1982), *Product Engineering Design Manual*, Kreiger, Malabar, FL.

Gregg L, and Steinberg E, (1980), *Cognitive Processes in Writing*, Lawrence Erlbaum and Associates, Hillsdale, NJ.

Günther J, Ehrlenspiel K, (1999), Comparing Designers From Practice and Designers with Systematic Design Education, *Design Studies*, 20(5), 439-451.

Günther J, Frankenburger E, and Auer P, (1996), Investigation of Individual and Team Design, in Cross N, Christiaans H, Dorst K, Eds. *Analyzing Design Activity*, John Wiley & Sons, 1996 Chichester, 117-132.

Harris G, (1997), Development of taste perception and appetite regulation, Bremner G, Slater A, Butterworth G, Eds. *Infant Development: Recent advances*, Psychology Press, Hove UK.

Harrison S, and Minneman S, (1996), A Bike in Hand: A study of 3-D objects in design, Cross N, Christiaans H, Dorst K, Eds. *Analyzing Design Activity*, John Wiley & Sons, 1996 Chichester, 417-436.

Hawkins J, (1974), Messing about in science, *The Informed Vision: Essay on learning and human nature*, Agathon Press, NY.

Heidegger M, (1962), *Being and Time*, translated by J. Macquarrie and E. Robinson, Harper and Row, NY.

Henderson K, (1999), *On Line and On Paper: Visual representations, visual culture and computer graphics in design engineering*, MIT Press, Cambridge, MA.

Hewitt M, (1985), Representational forms and modes of conception: An approach to the history of architectural drawing, *Journal of Architectural Education*, 39(2), (Winter), 2-9.

Holyoak K, Jun E, and Billman D, (1984), Development of analogical problem solving skill, *Society for Research in Child Development*, 55, 2042-2055.

Hubka V and Eder W, (1988), *The Theory of Technical Systems*, Springer-Verlag, Berlin.

Intons-Peterson M, (1993), Imagery and Classification, Collins A, Gathercole S, Conway M, Morris P, Eds. *Theories of Memory*, Lawrance Erlbaum and Associates, Hillsdale, 211-240.

Intons-Peterson M, (1993a), Imagery's role in creativity and discovery, Roskos-Ewoldsen B, Intons-Peterson M, and Anderson R, Eds. *Imagery, Creativity and Discovery: A cognitive perspective*, North-Holland Press, Amsterdam, 1-38.

Intons-Peterson M, (1996), Integrating the components of imagery, de Vega M, Intons-Peterson M, Johnson-Laird P, Denis M, Marchark M, Eds. *Models of Visuospatial Cognition*, Oxford University Press, 1996, 20-89.

Johnson-Laird P, (1996), Images, Models and Propositional Representations, in de Vega M, Intons-Peterson M, Johnson-Laird P, Denis M, Marchark M Eds. *Models of Visuospatial Cognition*, Oxford University Press, 1996, 90-127.

Johnson S, (1997), What's in a representation: why do we care and what does it mean?: Examining the evidence from psychology, *Representation and Design: ACADIA97 Conference Proceedings*, 5-16.

Kihlstrom J, (1996), Memory Research: the convergence of theory and practice, in Herrman D, McEvoy C, Hertzog C, Hertel P, and Johnson M, Eds. *Basic and Applied Memory Research: practical applications*, 1, Lawrence Erlbaum and Associates, Mahwah, NJ, 5-26.

Kosslyn S, (1980), *Image and the Mind*, Harvard University Press, Cambridge, MA.

Kosslyn, S, (1996), *Image and the Brain*, MIT Press, Cambridge, MA.

Kosslyn S, Brunn J, Cave K, Wallach R, (1984), Individual differences in mental imagery ability: A computational analysis, *Cognition,* 18, 195-243.

Krauss R, Myer J, Danielson S, and Lewis R, (1970), Design: a case history, in Moore G, ed, *Emerging Methods in Environmental Design and Planning*, MIT Press, Cambridge, MA, 11-20.

Langacker R, (1986), An Introduction to Cognitive Grammar, *Cognitive Science*, 10, 1-40.

Lawson B (1997), *How Designers Think: The Design Process Demystified*, Architectural Press, Oxford.

Likert R, and Quasha W, (1941), *Revised Minnesota Paper Form Board (Series AA)*, The Psychological Corporation, NY.

Lindemann P, and Wright C, (1998), Skill Acquisition and Plans for Actions: Learning to write with your other hand, in Osherson D, general editor, *An Invitation to Cognitive Science:* Scarborough D, and Sternberg S, eds. *Methods, Models and Conceptual Issues, 4*, MIT Press, Cambridge, MA, 523-584.

Lloyd P, Lawson B, Scott P, (1996), Can verbalization reveal design cognition? Cross N, Christiaans H, Dorst K, Eds. *Analyzing Design Activity*, John Wiley & Sons, 1996 Chichester, 437-461.

Maier NRF, (1931), Reasoning in humans II: The solution of a problem and its appearance in consciousness, *J. Comparative Psychology*, 12, 181-194; reprinted in Wason P, and Johnson-Laird P, (1968) *Thinking and Reasoning*, Penguin Modern Psychology, Baltimore MD, 17-27.

Mäntylä T, (1986), Optimizing cue effectiveness: recall of 500 and 600 incidentally learned words, *Journal of Experimental Psychology: Learning, Memory and Cognition*, 12, 66-71.

Massaro D, (1998), Models for reading letters and words, in Osherson D, general editor, *An Invitation to Cognitive Science:* Scarborough D, and Sternberg S, eds. *Methods, Models and Conceptual Issues, 4*, MIT Press, Cambridge, MA, 301-364.

Miller A, (1984), *Imagery in Scientific Thought: Creating 20th century physics*, Birkhauser, Boston.

Newell A, (1969), Heuristic Programming: Ill-structured Problems, in Julius Aronofsky, ed. *Progress in Operations Research*, 3, John Wiley & Sons, 360-414.

Newell A, and Simon H, (1972), *Human Problem Solving*, Prentice Hall, Englewood Cliffs, NJ.

Paivio A, (1971), *Imagery and Verbal Processes*, Holt, Rinehart and Winston, NY.

Paivio A, (1986), *Mental Representations: A dual coding approach*, Oxford University Press, UK.

Pahl G, and Beitz W, (1996), *Engineering Design*, Springer-Verlag, London.

Peterson M, (1993), The ambiguity of mental images: insights regarding the structure of shape memory and its function in creativity, in Roskos-Ewoldsen, Intons-Peterson M, and Anderson R Eds. *Imagery, Creativity and Discovery: a cognitive approach*, Elsevier, Amsterdam, 151-186.

Purcell T, Ed. (1998), Special Issue on Sketching and Drawing in Design, *Design Studies*, 19(4).

Purcell T, Gero J, et al. (1996), The Data in Design Protocols: The Issue of Data Coding, Data Analysis in the Development of Models of the Design Process, Cross N, Christiaans H, Dorst K, Eds. *Analyzing Design Activity*, John Wiley & Sons, 1996 Chichester, 225-252.

Pylshyn Z, (1973), What the mind's eye tells the mind's brain: a critique of mental imagery, *Psychological Bulletin*, 80, 1-24.

Radcliffe D, (1996), Concurrency of actions, ideas and knowledge displays within a design team, in Cross N, Christiaans H, Dorst K, Eds. *Analyzing Design Activity*, John Wiley & Sons, 1996 Chichester, 343-364.

Reisberg D, and Logie R, (1993), The ins and outs of working memory: Overcoming the limits on learning from imagery, Roskos-Ewoldsen, Intons-Peterson M, and Anderson R Eds. *Imagery, Creativity and Discovery: a cognitive approach*, Elsevier, Amsterdam, 39-76.

Reitman W, (1964), Heuristic Decision Procedures, Open Constraints, and the Structure of Ill-defined Problems, Shelly M, and Bryan G, Eds. *Human Judgments and Optimality*, Wiley, New York, NY, 282-315.

Robbins E, (1994), *Why Architects Draw*, MIT Press, Cambridge, MA.

Sachs A, (1999), 'Stuckness' in the design studio, *Design Studies*, 20(2), 195-210.

Schon D, (1983), *The Reflective Practitioner*, Basic Books, NY.

Schon D, (1988), Designing: Rules, types and worlds, *Design Studies*, 9(3), 181-190.

Schon D, (1992), Designing as Reflective Conversation with the Materials of a Design Situation, *Research in Engineering* Design, 3, 131-147.

Schon D and Wiggins G, (1992), Kinds of seeing and their functions in designing, *Design Studies*, 13(2), 135-156.

Seel N, and Dörr G, (1994), The supplementation of mental images through graphics: instructional effects on spatial visualization skills of adults, Schnotz W and Kulhavy R, Eds. *Comprehension in Graphics*, North-Holland Press, Amsterdam, 271-290.

Shepard R, and Feng C, (1972), A Chronometric Study of Mental Paper Folding, *Cognitive Psychology*, 3, 228-243.

Shepard R, (1978), *Mind Sights*, Freeman, NY.

Shepp B, (1989), On perceiving objects: holistic versus featural properties, Shepp B, and Balllesteros S, Eds. *Object Perception: Structure and Process*, Lawrence Erlbaum and Associates, NJ, 203-233.

Simon H, (1969), *The Sciences of the Artificial*, MIT Press, Cambridge, MA.

Simon H, (1973), The structure of ill-structured problems, *Artificial Intelligence*, 4, 181-203.

Simon H, (1976), The information storage system called 'Human Memory' in Rosenzweig M, and Bennett E, Eds. *Neural Mechanisms of Learning and Memory*, MIT Press, Cambridge, MA, 79-96.

Suh N, (1990), *The Principles of Design*, Oxford University Press, Oxford.

Tafel E, (1979), *Apprentice to Genius: Years with Frank Lloyd Wright*, McGraw-Hill Book Co.

Takeda H, Yoshioka M, Tomiyama T, and Shimomura Y, (1996), Analysis of Design Protocol by Functional Evolution Process Model, Cross N, Christiaans H, Dorst K, Eds. *Analyzing Design Activity*, John Wiley & Sons, 1996 Chichester, 187-210.

Talmy L (1988), Force Dynamics in Language and Cognition, *Cognitive Sciences,* 12, 49-100.

Trousse B, and Christiaans H, (1996), Design as a Topos-based Argumentative Activity: A protocol analysis study, Cross N, Christiaans H, Dorst K, Eds. *Analyzing Design Activity*, John Wiley & Sons, 1996 Chichester, 365-388.

Ullman D, (1992), *The Mechanical Design Process*, McGraw-Hill, NY.

Ullman D, Herling D and Sinton A, (1996), Analysis of Protocol Data to Identify Product Information Evolution and Decisionmaking Process, in Cross N, Christiaans H, Dorst K, eds. *Analyzing Design Activity*, John Wiley & Sons, 1996 Chichester, 169-185.

Ulusoy Z, (1999), To design versus to understand design: the role of graphic representations and verbal expressions, *Design Studies,* 20(2), 123-130.

Von der Weth R, (1999), Design Instinct? — the development of individual strategies, *Design Studies* 20(3), 453-463.

Visser W, (1996), Use of Episodic Knowledge and Information in Design Problem Solving, in Cross N, Christiaans H, Dorst K, Eds. *Analyzing Design Activity*, John Wiley & Sons, 1996 Chichester, 271-290.

Viviana P, and Cenzato M, (1985), Segmentation and coupling in complex movements, *J. Experimental Psychology: Human Perception and Performance,* 11, 828-845.

Vosniadou S, (1989), Similarity and analogical reasoning: a developmental perspective, in Vosniadou S, and Ortnony A, Eds. *Similarity and Analogical Reasoning*, Cambridge University Press, Cambridge, 199-239.

Vosniadou S, and Ortnony A, Eds. (1989), *Similarity and Analogical Reasoning*, Cambridge University Press, Cambridge, MA.

Winograd T, and Flores F, (1987), *Understanding Computers and Cognition: A new foundation for design*, Ablex Corporation, Norwood, NJ.

Wallace B, and Hofelich B, (1992), Process generalization and the prediction of performance on mental imagery tasks, *Memory and Cognition*, 20, 695-704.

Ward T, Finke R and Smith S, (1995), *Creativity and the Mind: discovering the genius within*, Plenum Press, NY.

Wright F L, (1928), Concept and Plan, *The Architectural Record*, January-February.

Chapter 9

Visual Analogy – a Strategy for Design Reasoning and Learning

Gabriela Goldschmidt
Technion – Israel Institute of Technology
Haifa, Israel

Abstract

This paper proposes that the use of visual analogy in problem solving is an example of similarity-based reasoning, cognitively facilitated by imagistic operations. We show that in designing, which is an example of ill-structured problem solving, this type of reasoning is most valuable. We offer evidence that novices, in particular, benefit from guidance to use analogy, which helps them to better understand abstract concepts and to fully exploit their capacity to retrieve and implement previously acquired knowledge.

1 Two Systems of Reasoning

Four decades ago the design research community in its entirety, more or less, subscribed to the notion that design practices should be made to follow exacting methods, the development of which became the single most important concern on design research agendas. Members of the so-called "design methods movement" worked hard to help develop a design science: it was seen as an important step forward in the evolution of the design disciplines and their standing among the professions. (The professions, in turn, struggled to gain an academic status that would put them on an equal footing with the sciences). Consequently, so-called 'traditional' design practices, based on 'soft' strategies including intuition and the use of informal knowledge were looked down upon. In contrast, the idea of a 'hard' analytic science of design sounded very appealing, as eloquently expressed by members of the methods movement (e.g. Gregory 66) and by Herbert Simon, in *The Sciences of the Artificial* (Simon 81).

The experience of the last few decades has taught us that we are far from attaining the goal of formulating a design science; indeed, the very goal has been submitted to serious questioning. It turns out that in most if not all design domains, rigorous design methods based on well-defined algorithms do not yield the expected improvement in design quality. In some cases, particularly in architecture and in industrial design, it is very hard to get designers in the real world to even try them out. As a result, we have been forced back to the drafting board – and to the keyboard as well, this time – to take a hard look at how designs, and primarily design ideas, are being generated. Two interrelated questions beg to be answered: why should design methods, based on carefully articulated sets of rules, fail to deliver better designs? And why do such methods meet with considerable resistance on the part of so many design practitioners who refuse to give up their age-old 'soft' design strategies, despite their inferior status as compared to 'scientific methods'? These questions are especially important from the perspective of design education. On the one hand, we are uncomfortable with our traditional, experiential, 'trial and error' teaching traditions; on the other hand, we fall short of being able to effectively transmit a comprehensive codex of design knowledge, including procedural knowledge about how to design and *how to reason* about designing. It is for this reason that design reasoning has become a research topic of great importance, motivated by the hope that knowing more about natural design reasoning processes would pave the way for modes of enhancing and teaching it. Fortunately, this question falls within a more general research agenda on human cognition, in the area of reasoning and problem solving, and work in this area is useful for our quest.

For a long time psychologists took it for granted that development follows a trajectory leading from reasoning based on similarity in early childhood, to rule-based reasoning in adulthood. The latter, in this view, signifies the ability to form abstract concepts and use symbolic representation, primarily through the use of language. This belief was, and still is in some quarters, so fundamental that very few researchers have challenged it. Similarity pertains to figurative qualities, and is therefore largely visual. Fodor (75), talking about the nature of mental visual imagery, has simply stated that children think in pictures, whereas adults think in words. Kaufmann (80), who thought it wrong but who found himself fighting windmills, has talked about a "sweeping linguistic imperialism" (p. 18). According to him, linguistic representation is used to reason about problems that belong to a familiar type. For novel problems he thinks imagery is enlisted, and the reason imagery is helpful in those cases is the relative freedom of imagery from rules.

In later years cognitive scientists talked about "pop-up" or bottom-up cognitive operations in reasoning, roughly parallel to similarity and example-based inductive reasoning, as more rudimentary than the more sophisticated, central executive bottom-down cognitive operations which correspond (roughly, again) to rule-based deductive reasoning.

In recent years the primacy of rule-based reasoning is no longer universally accepted as an absolute truth, and researchers are interested in the relationship between the two modes of reasoning: rule based and similarity based, in both children and adults (Sloman and Rips 98). Sloman (96) has suggested that we are endowed with two independent (but interacting) cognitive systems, each dedicated to one mode of reasoning: one associative and similarity-based, the other symbolic and rule-based. In Sloman's account, however, there still is a clear hierarchy and the associative system is seen as 'subsymbolic'. Other researchers have advanced the view that if indeed there are two systems of reasoning, they are equally important to processes of problem solving and learning (i.e. Gentner and Medina 98). A wider discussion of this most interesting question is unfortunately beyond the scope of the present work. For our purposes here suffice it to say that we accept Gentner and Medina's view and our arguments will proceed on the assumption that two parallel, interactive cognitive systems operate in problem solving, one underlying reasoning on the basis of rules, and the other on the basis of similarity.

To the two-fold reasoning paradigm we add an assumption of our own, which we do not attempt to prove here, as following: The strength of activation of one or the other system of reasoning in problem solving is context dependent. There are individual differences among problem-solvers (and developmental differences as well), but mostly the type of problem, and especially whether it is well-structured or ill-structured, determines which reasoning system is most instrumental for a particular instance of problem solving. Furthermore, we postulate that for ill-structured problems similarity-based reasoning is especially useful, as there are few or no rules the problem-solver can retrieve and use (one of the properties of ill-structured problems is that no means are specified for reaching a problem's end-state, or solution, from the initial state, through interim states). Since we see designing as a process of solving problems that for the most part are ill-structured, it will come as no surprise that we are particularly interested in similarity-based reasoning.

2 Visual Analogy: An Imagistic, Similarity-Based Reasoning Strategy for Ill-Structured Problem Solving

Donald Schön, who was interested in the formulation of problems as the basis for action throughout his career, pointed out decades ago (Schön 63) how powerful the transfer of concepts from one case to the next and from one field to another can be. Gentner and Medina (98) have presented rich evidence to support the claim that similarity-based reasoning is a powerful engine in the solving of many types of problems. Of the various kinds of similarities that may be activated in a process of mapping and transfer from one situation to another they single out analogy. Analogy, which refers to similarity between relationships, is of particular importance because in its most complete state it establishes structural commonalties, in addition to surface features that are carried over to a new situation by other types of similarity as well[1]. It is true that the carrying over of surface

[1] We would like to stress the importance of the definition of analogy as similarity between relationships, which distinguishes it from other kinds of similarity. Analogy is sometimes equated to metaphor and even confounded with it. In our view, analogy and metaphor are two very different types of similes.

features only, without a structural similarity to underpin them, may lead to a false analogy and consequently to a wrong solution to a problem. We have discussed and illustrated this problem elsewhere (Goldschmidt 95). However, we must qualify this assertion when talking about visual analogy, where transfer of features that may be superficial otherwise, like form and shape, are actually the essential core of a situation. It is therefore important to exercise caution when determining which properties of a problem are structural and which are superficial, as their definition may vary according to the type of problem. Design problems are often concerned with form giving and therefore this question is of particular relevance to us. Gentner and Medina (98) have also shown that the use of analogy, especially by novices, contributes to the learning of new abstract concepts and to the extension and application of previously acquired knowledge. We will return to the question of learning later.

If the use of analogy does indeed facilitate problem solving of a certain type, and we have assumed that these problems are primarily ill-structured, then we must address the hitherto unanswered question: what are the underlying cognitive mechanisms that are involved in the generation of analogy, and visual analogy in particular? An answer to this question may help explain why analogy is rarely used spontaneously, as suggested by the results of several studies (e.g. Beveridge and Parkins 87; Gick and Holyoak 80; 83. Most studies styled their experiments after Duncker's famous radiation problem (Duncker 45)). [2] It may also help decide if and when it makes sense to induce the use of analogy as a problem solving strategy, especially in educational settings. We can treat these questions only in a preliminary and partial manner in this paper, and we hope that further research will make it possible to elaborate on these issues. The cognitive mechanism that we hold 'responsible' for visual analogy, which is most instrumental in design problem solving, is mental visual imagery. In the following sections we discuss the relationship between visual analogy and imagery in creative problem solving as we understand it, to be followed, in section 4, by commentary on spontaneity or lack thereof, and on educational benefits ensued from the use of guided analogy as a design strategy.

2.1 Kekulé's Insight

"...Again the atoms were gamboling before my eyes. This time the smaller groups kept modestly to the background. My mental eye, rendered more acute by repeated

[2] In this problem subjects were asked to suggest an appropriate medical procedure for a patient who suffers from a tumer in his stomach. The tumer is not operable, but a certain kind of radiation can destroy it; the problem is that at the high intensity that is required, helthy tissue around the tumer would also be destroyed. Before they were given this problem, subjects were told a story about a general who wanted to capture a fortress; this was not possible in a frontal attack because of mined roads; the solution was to divide the force and have the soldiers attack in small groups from several directions at once.

vision of this kind, could now distinguish larger structures, of manifold conformation; long rows, sometimes more closely fitted together; all twining and twisting in snakelike motion. But look! What was that? One of the snakes had seized hold of its own tail, and the form whirled mockingly before my eyes. As if by a flash of lightning I awoke..."

This well-known citation, attributed to the 19th century chemist Friedrich Kekulé (e.g. Koestler 64, 118; Rothenberg 79, 396), is an introspective account of what went on in his mind when he had an insight that eventually led to the discovery of the ring structure of benzene (see Figure 1). The authenticity of the actual event is questionable: Kekulé made the famous comments about his insight decades after the event, in circumstances that suggest that he may have wished to be entertaining rather than factually accurate (Gruber 95). The fact that the insight allegedly occurred while Kekulé was half-asleep or at least engaged in day dreaming also adds to doubts about the scientist's ability to give a credible account of his thinking. Some critics go as far as to argue that the entire story amounts to nothing less than a monumental case of scientific fraud (Wotiz, Rudofsky 84; Wotiz 93). We are not in a position to comment on the authenticity question, nor does it make much difference for our purpose whether the insight actually did take place, and under what circumstances. Since the story is conceivable and resembles countless other cases of anecdotal evidence regarding insights that were instrumental in creative acts of problem solving, we shall accept it on the basis of its plausibility.

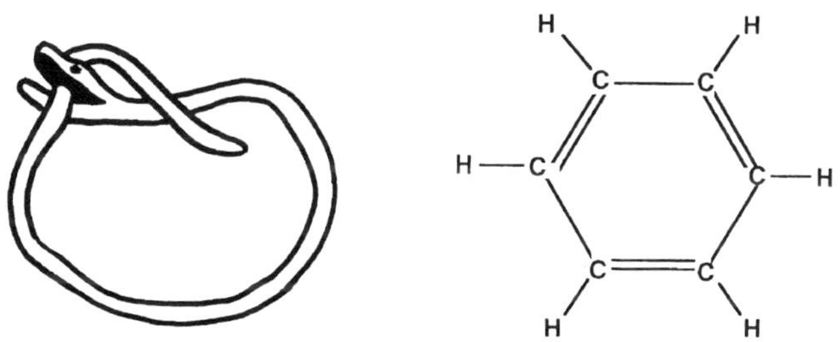

Figure 1: Kekule's insight and discovery of the structure of the benzene molecule

The story is repeated in many scientific discussions and is used to illustrate theories of scientific discovery. Interestingly, different authors utilize the Kekulé story to advance different arguments. In particular, it serves as either an example of mental visual imagery (e.g. Finke 90; Miller 86; Perkins 81; Shepard 78), or as an instance of analogical mapping and transfer (e.g. Ariety 76; Holyoak, Thagard 95; Langley, Jones 88; Martindale 95; Peterson 93). Robert Weisberg, a prominent psychologist, appears to be undecided, or else he sees no problem in citing the case alternately as an example of imagistic reasoning on one occasion, and analogical reasoning at another opportunity (Weisberg 86, 88).

Proposition: The dual significance of the Kekulé insight, or shall we say anecdote, helps bring forth the question we are concerned with: what is the relationship between visual imagery and visual analogy? In particular, we are interested in the use of imagery and analogy in reasoning, especially in processes of creative problem solving. Are imagistic reasoning and analogical reasoning two separate processes, that may or may not occur in tandem? If Kekulé's case can indeed be seen as both a case of imagery and a case of analogy, there are three possible relationships among these two cognitive phenomena, leading to the following alternative propositions: a) two distinct operations occurred in sequence: first imagery, then analogy b) two distinct operations coincided: systems that activate imagery and systems that activate analogy operated concurrently. c) a single imagistic system operated; one or more subsystems activated imagery in the specific tasks of mapping and transfer that we refer to as visual analogy. We propose that the third proposition is most convincing, provided we accept a wide enough definition of imagery. We shall substantiate this claim through the use of an example from the domain of design, where both imagistic reasoning and analogical reasoning are quite frequent. [3]

2.2 The Dommelsch Advertisement Campaign

Nico van Scherpenzeel is a senior art director in a large Dutch advertising company that was commissioned to design an advertisement campaign for Dommelsch beer. It was to be multi-functional (posters and magazines, television and radio. Here we focus on the posters only) and was meant for the summer months (91). The design process description that follows is based on several conversations with van Scherpenzeel, who was in charge of the endeavor, approximately six months after the campaign was launched.

The design process can be divided into four phases. In phase 1 the designers looked for a theme. After a long search during which many ideas were considered and discarded, they decided to capitalize on the fact that it was summer time and chose to compare having Dommelsch beer with going abroad on vacation. The theme was encapsulated in the slogan: "*But we have Dommelsch.*" van Scherpenzeel testifies that this decision and the slogan that goes with it came in a flash. In phase 2 a prototype poster was developed, envisaged as containing a picture (photograph) of a desirable item, such as a beautiful car or house, next to which (to the right) the thematic slogan was to be printed out. Phase 3 still contained one

[3] We do not yet possess empirical evidence to support our claim that visual analogy is a special case of visual imagery; however, it should be possible to run neurocognitive experiments using fMRI, PET or other imaging devices, to inspect the respective areas of the brain that are activated during visual analogy and visual imagery processes.

picture, but instead of just any item, the picture was to show a famous building or site in a popular vacation destination. The slogan was now to be printed below the picture. In phase 4 the final idea was crystallized: van Scherpenzeel refers to it as *picture-rhyme*. The single picture was replaced by two "rhyming" pictures: one depicting the tourist attraction (as in phase 3), and another (to the right, or beneath) showing Dommelsch beer (bottle, glass, crate, etc.). The slogan is added as before.

Figure 2: The 'Great Wall of China' Dommelsch advertisement [4]

First three posters were produced, to be followed later by three more. The production of the

[4] Advertisement by Noordervliet and Winninghoff/Leo Burnett, Amsterdam (Nico van Scherpenzeel, Art Director) for Dommelsch Bierbrouwerij B.V. Reproduced with permission.

series followed the usual working procedure in advertisement: the site-photographs were mostly stock items supplied by an agency according to specifications that took into account the resemblance to a Dommelsch bottle (or glass, etc.). The sites included the Statue of Liberty, the guards of Buckingham Palace, the tower of Pisa, the Eiffel tower, the Niagara Falls and the Great Wall of China. We will use the latter poster, shown in Figure 2, as an exemplar. First an attractive photo of a section of the China Wall was chosen. Then a complicated operation took place to fabricate its Dommelsch analog: two photos were taken: one of a flat landscape, the other of a configuration of beer crates, on a flat surface. The two images were fed into a computer where they were combined and manipulated to create a composite image. In the final picture beer crates appear to be arranged in a hilly landscape, such that the geometry of both the beer crate construction and the landscape features bear a striking resemblance to the Wall of China and the landscape in which it is set. The China photo was selected from amongst several photos of the Wall that were supplied for inspection as per request. van Scherpenzeel testifies that the idea of the crate-arrangement flashed the minute he had first contemplated the particular photo that was subsequently selected.

2.3 Process Reconstruction and Analysis

What does the story of the Dommelsch poster design have in common with the insight that led to the discovery of the benzene molecule structure? We propose that the inventive processes in the two cases (different as they might be in magnitude and significance) were similarly served by analogical reasoning executed through mechanisms of imagery. In both cases the individuals involved had been working on their respective problems for some time: deciphering the structure of a molecule, and designing an effective advertisement campaign for beer. Both men were experienced thinkers in their domains. Kekulé had of course dealt with molecule-structures for a long time, and had considered many alternative configurations for the benzene molecule, before and after his insight. Hexagonal configurations in plane were one such possibility (Gruber 95). Kekulé used imagistic reasoning as a matter of routine; in fact he "had developed a special form of imagery in which he could think directly; he had already had many ideas in that modality, and now he had a new one." (Gruber ibid. 410). Likewise, van Scherpenzeel was no newcomer to the design of advertisement campaigns, and by the time he was ready for the last poster in the Dommelsch series, he had already successfully established the "picture-rhyme" strategy. Both men were committed to their work, and thought about their tasks continuously, conducting active searches for possible solutions while using their personal versions of the tools of their trades.

The tools of the trade deserve our special attention, because they include consulting pictorial material of various kinds. van Scherpenzeel retrieved photographs related to the task from an external database, although he did not and could not know how many pictures he might have to inspect before one that could be of use would be found. Kekulé, who did not have a database at his disposal, nor had he advanced his search to the point where he could explicitly specify what type of pictures he would be interested in inspecting, resorted to the "special form of imagery" that he was practiced in. He knew from past experience that if he allowed himself to generate seemingly random images, the preoccupation with a specific problem would most probably place constraints on those images such that only relevant images would be formed. Alternatively, relevant images would attract his attention sufficiently to notice their relevance (compare to Chase and Simon's findings that expert chess players avoid trial and error and to a large extent are able to make successful moves only) (Chase, Simon 73). Kekulé's account contains no information regarding any other images he might have entertained during the long search related to his problem solving efforts. It is safe to assume, however, that he had had other images that may have faded away leaving no trace because he could not identify any helpful clues in them, hence expelling them to the realm of irrelevancy. Thus both individuals inspected pictures, externally or internally, with the explicit intention of finding in them hints that could potentially advance their search efforts. (Compare these 'hunting' episodes with a report by de Bono on the development of a medical device to test lung function. de Bono deliberately decided to wander around a large dime-store because it was "a setting full of the sort of objects which might trigger an idea. It was just a matter of wandering around until something suggested itself" (de Bono 71, 111).

Relevant pictures were found. Kekulé "saw" twisting snakelike rows, one of which "had seized hold of its own tail", thus forming a ring. van Scherpenzeel saw a vivid composition depicting three stretches of the wall with three watch-towers, placed where the wall changes direction. Why were these pictures useful? Because our problem-solvers could form images of analogous pictures, that provided the basis for solutions to their problems. They did not simply *identify* analogies: they *created* them by imaging a target configuration. In Kekulé's case - a ring of carbon atoms; in van Scherpenzeel's case - a configuration of beer bottle crates laid out in a landscape. Kekulé's target image was a model that needed testing and evaluating. van Scherpenzeel's target image had to be fabricated in reality to resemble the inner picture (using photographic and computational techniques). With hindsight it is easy to identify the analogy between the source and target pictures in each case. But the processes of mapping and transfer from source to target are not so obvious and indeed, they must have been two-way processes, such that source to target moves were accompanied by target to source moves as well. In other words, the source is not merely retrieved from memory; it may be retrieved and transformed, or actually reconstructed and reorganized almost from scratch until its representation is suggestive enough so that mapping and transfer to a candidate target that is being constructed becomes possible [5]. In this view, neither source nor target are stable

[5] A case in point that illustrates the role of transformation in retrieved information is the erroneous representation of a famous building plan, in the process of resolving an architectural problem. The

representations until the process of transfer has been completed: both representations are susceptible to transformations in order to strengthen and verify the accuracy of mapping.

Two points must be emphasized. First, tasks that are aimed at creating novel objects and entities (two and three-dimensional), such as various design assignments or scientific modeling that is expressed in diagrams etc., naturally require visual or pictorial representations. Inner representations through imagery are known to complement external representations of various sorts. It is therefore not surprising that individuals who solve problems of this type make use, extensive use at times, of their imagistic powers. Second, we must remember that there are differences between well-structured problems and ill-structured problems, such as the design and scientific discovery problems we are concerned with. In solving the latter a search is indispensable, as most ill-structured problems have more than one possible solution. When an ill-structured problem also falls into the 'visual' category, then all kinds of configurations may provide bases for a representation that would be considered a satisficing (to use Herbert Simon's terminology) solution to the problem. Using all available strategies, the problem solver is likely to conduct a search in which pictorial representations, external and internal, are generated and transformed.

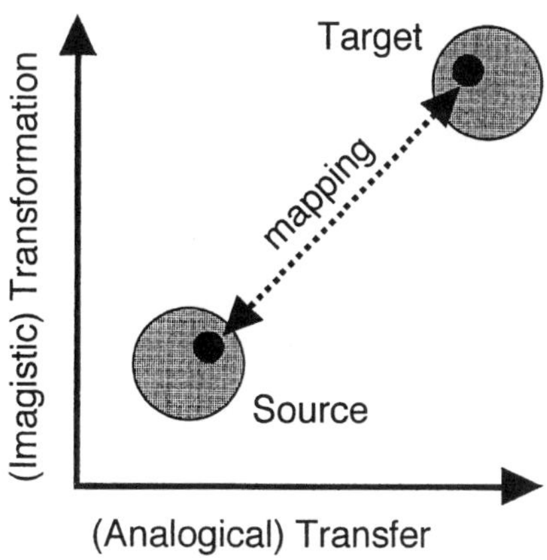

Figure 3: Diagram of dynamic imagistic/analogical search process in creative problem solving

This explains the importance of the dynamic principle we outlined above, which allows for the entertainment of unstable, possibly vague images, that may undergo serious

---

error permitted the architect to form a visual analogy that helped him resolve his problem (Goldschmidt, 1994; 1995).

transformations throughout the process of transfer between candidate-source and candidate-target. Figure 3 is a diagram of the dynamic, two-way mapping process in a creative search involving imagistic transformation and analogical transfer.

3 Interactive Imagery

If image generation is a very complex cognitive phenomenon, imagistic reasoning must involve even more complex cognitive structures. Kosslyn (94) counts four image generation types classified according to the mechanisms that trigger them, and differentiating between arrangements of image components and the individual components themselves. Image components can be arranged using either categorical or coordinate representations of spatial relations, and individual components can be imaged by allocating attention or by retrieving and activating visual memories. In imagistic processes involved in problem solving such as we have described, images are formed that were never before perceived: the designer, the inventor and the discoverer work to represent something new, a brainchild that has never previously been stored in memory and can therefore not be retrieved. Kosslyn explains this possibility as follows: "People can "mentally draw," visualizing patterns that they have never seen and that are not simply new combinations of familiar components. This process is a special case of the attention-based imagery..." (Kosslyn 94, 310). Elsewhere Kosslyn adds that this special type of imagery utilizes image transformation, i.e. the ability to modify patterns in images (Kosslyn 99). Imagery based on attention implies interpretative processes and indeed, Finke (90) has shown that by manipulating given components (selected randomly from a restricted set) in imagery, people can successfully combine them into useful objects when interpretative constraints are imposed (the object is to belong to a specified category, say furniture). Finke found that not only could people perform such tasks fast and arrive at results, but that the exercise yielded unique objects that were often rated as creative by judges who evaluated them. So convincing were the findings regarding the power of imagery, that Helstrup and Anderson set forth to show that various decomposition tasks can be performed in imagery alone as accurately as when subjects were also allowed to sketch in the process (Helstrup and Anderson 91).

However, in Helstrup and Anderson (ibid.) and in subsequent studies of mental synthesis, sketching experience was not a variable and subjects where not experienced sketchers. Versijnen et al. (98), who were familiar with designers' propensity for sketching, found those results counter intuitive. They were able to show that in composition tasks that resemble Finke's experiments, creativity scores favor sketchers over subjects who used only imagery, when the subjects were experienced sketchers and when their manipulation of the given components included restructuring. When the subjects were not experienced sketchers, no significant differences in creativity were found between those who sketched and those who used only imagery, as in Helstrup and Anderson (ibid.). Sketching skills are therefore belived to have the power to amplify imagery, when exercised.

We relate these results here because they appear to support the claim that imagery can be 'stretched'. In other words, depending on task requirements and an individual's abilities, a larger or smaller number of imagistic systems or subsystems may be activated. If a composition task can benefit from external representation and if the individual is fluent in the production of such representations (no cost or low cost in terms of attention), then we can expect sketching to be a contributing factor. This is certainly true for design tasks in the hands of skilled designers. Goldschmidt (91) has called the act of sketching in the early phases of conceptual design 'interactive imagery', meaning "the simultaneous or almost simultaneous production of a display and the generation of an image that it triggers." (ibid. 131). Interactive imagery enables the designer to converse with his materials, to use Schön's terminology, in a dynamic manner, taking advantage of the speed with which images and sketches can be generated and transformed.

In some respects the use of visual analogy in creative processes serves a purpose similar to that of sketching: a problem-solver faced with a complex ill-structured problem, in a domain that requires visual representation, recruits available resources that can fortify his or her search strategies. Since no rules are at hand to reason with, the immediate goal is to generate visual representations that are rich with clues that may potentially lead to a solution by way of similarity. There are differences between the use of an external support tool such as sketching and the spontaneous (for the most part, but not always) search for and identification of analogous situations, which is not necessarily accompanied by an external representation. But there is no contradiction between these strategies which are not mutually exclusive, and may actually be practiced concurrently, as the representations in question are external and internal respectively and orthogonal to one another.

Sketching can, of course, be practiced independent of imagery (and for purposes other than inventive problem solving). Likewise, an analogy can be identified without the use of imagery. However, in a creative search analogies are not just identified, they are created as a result of the manipulation and transformation of images, as we have seen earlier. Outside of the laboratory, it is very rare for a problem-solver to be presented with two displays that can be identified as analogous. It is almost always the case that the problem-solver contemplates one display (externally or internally), and forms an image of another display that is identified as analogous to the first. As noted earlier, it is of no consequence which of these displays is the source and which is the target. Images undergo transformations, because the problem-solver uses them to reason about the problem at hand. The transformations bring the two images close enough together to allow mapping and transfer, and this is true whether we subscribe to a theory of structural relations or a theory of functional multiconstraints in analogy (e.g. Gentner 83; Gick and Holyoak 80; Holyoak, Thagard 95; Keane 88). These are the arguments that drive us to believe that visual analogy is a cognitive operation that can only occur when imagery is activated. Seeing it as a subsystem of imagistic mechanisms makes it possible to posit that the analogy subsystem is activated when a complex problem requires amplification of the basic imagistic system. It is reasonable to assume that this subsystem is more developed in some individuals, who are therefore likely to exploit it more effectively. This explains the relative rareness of spontaneous appeal to analogy in general problem solving, and its successful utilization by certain specialists.

Like imagistic reasoning, which turns out to be an elaborate ability in which we can identify a number of types, analogical reasoning is also a complex cognitive capability. We distinguish between surface and deep analogies, and between within-domain and between-domain analogies or, to be even more elaborate, between local, regional and long-distance analogies (Dunbar 95). Due to historical circumstances, the study of imagery and analogy have developed as two distinct research fields that have little in common. This may be in part due to the fact that analogy researchers seem to have taken for granted the fact that so many instances of analogy involve visual representations. In reported instances of the use of analogy in creative problem solving, researchers remain indifferent to the question of whether these representations are external or internal. Elsewhere, the effort to model analogical reasoning computationally has concentrated on the identification of analogies rather than on their generation. The primarily visual nature of the information to undergo mapping and transfer, and the fact that it is not always simply perceived but may actually have to be generated first, have never received ample attention (e.g. Vosniadou and Ortony 89).

4 Visual Analogy and the Acquisition of Design Expertise

While we know what a decisive role analogical reasoning has played in many a discovery, invention and creative achievement, we also have substantial evidence that spontaneous recourse to analogy as a reasoning strategy in problem solving is quite uncommon, to say the least. Many known instances of beneficial use of analogy involve visual images, and we may therefore assume that the habit of "thinking in pictures" (in adulthood) may actually lead to a better chance of employing analogical reasoning in problem solving. It is therefore reasonable to hypothesize that problem-solvers will perform better if made to use visual images, and in particular, if made to use analogy when it is doubtful that they would do so spontaneously. Should this prove to be the case, we may want to ask what implications for learning in general does the induced use of analogy have, if any? And what specific implications can we identify for learning to design?

Schunn and Dunbar (96) have shown that in science, priming enhances access to previously acquired knowledge, which can be and is used to solve current problems, among others through analogical mapping and transfer. This process is implicit and in an experiment, subjects were not aware of what made them think of the particular knowledge they were able to elicit in order to successfully solve a novel problem. As we have already reported above, Gentner and Medina (98) who conducted experiments involving the use of analogy with children, have come to comparable conclusions. They stress that not only do subjects perform better when they had previously solved analogous problems, but that long-term learning can be shown to occur in which abstract concepts are better understood, abstracted, and applied in subsequent appropriate situations, (Note the difference between these results and those obtained by Gick and Holyoak, where problem solving was not enhanced following passive exposure of subjects to analogous situations. See section 2).

Casakin and Goldschmidt (Casakin 97; Casakin and Goldschmidt 99, Casakin and Goldschmidt 00) asked whether the use of analogy would produce similar effects in design problem solving, and set out to investigate this question empirically. Three groups of subjects with varying design experience participated in the experiments: experienced architects, advanced architecture students and novice architecture students. Each subject solved a number of well-defined (one correct solution) and ill-defined (any number of acceptable solutions) design problems that were presented under three conditions: a) The design problem was administered with no additional material or commentary; b) The design problem was administered while the subject was also shown a panel with visual displays (approximately two dozen images), some of which could be used as source analogs for the problem, and some that could not be related to it in any way. The latter served as 'noise' [6]; c) Same as b, but subjects were also explicitly encouraged to identify relevant images among the displays and use them as source analogs in their designing. The subjects' solutions to the design problems were expressed in sketch form, on one or more sheets of paper. These sketch-designs were assessed by three naïve judges (all experienced architects) for arrival at a solution in the case of well-defined problems (yes/no scores), and for the quality of the solutions and their creative merits on a scale of 5 points in the case of ill-defined problems (for the purpose of this study we do not differentiate between well/ill-defined, and well/ill-structured problems). As the degree of agreement among the judges was very high, the results were seen as perfectly reliable and were consequently submitted to statistical analyses. Details of the experimental settings and the results are reported elsewhere (Casakin, ibid; Casakin and Goldschmidt, ibid.); here we want to highlight some of the findings which, we believe, are highly relevant to our study of analogy as a reasoning strategy in designing, and in learning to design.

---

[6] For example, one of the problems called for the design of a typical floor of a prison, such that the guards maintain visual control over the doors of all 80 cells; each cell was to have at least one exterior wall. Some of the images in the panel of displays that accompanied the presentation of this problem described the following: a wheel (with spokes); a clock, a top view of a round slide projector carousel; the logo of the Shell company; a fan; a plan of a fortified city; a number of floor plans of residential buildings with a central core (elevators etc.) and radial 'fingers' in which apartments are located.

We choose to distinguish between experimental results obtained for problems termed 'well-defined' and those termed 'ill-defined'. For well-defined problems, architects were assisted by the presence of visual displays but instructions to use analogy had no significant effect on the quality of their work. In contrast, students' performance was enhanced by instructions to use analogy, whereas the mere presence of visual displays had no significant improving effect on their work. For ill-defined problems, on the other hand, the pattern of results was different. Both architects and students obtained significantly higher scores when using visual displays, and giving them instructions to use analogy maximized their design-quality scores. These results are summarized in Table One (based on findings reported in Casakin 97. Analysis in Casakin and Goldschmidt 99 and Casakin, Goldschmidt 00 pertains only to ill-defined problems). Our interpretation of these findings follows below.

Table 1: Design quality scores, novice and experienced designers under different problem solving conditions

		Novice designers (students)				Expert designers (architects)	
		Well-defined		Ill-defined		Well-defined	Ill-defined
		Begin.	Advan.	Begin.	Advan.		
Baseline (control)	No display provided No instructions to use analogy	--	--	2.227	2.58	--	2.809
Design quality improvement	Displays provided No instructions to use analogy	p<.469	p<.541	2.621 p<.039	2.939 p<.092	p<.091	3.236 p<.046
Additional design quality improvement	Displays provided Instructions to use analogy given	p<.063	p<.042	3.463 p<.001	3.731 p<.002	p<.148	3.984 p<.001

For experienced professionals, well-defined problems are not novel and the architects in our experiments could solve them using routine, low-level strategies. The visual displays they contemplated, when they did, where enough to evoke similar problems they had solved in the past, and possibly even straight-forward algorithms they could easily access and activate for a successful solution of these problems. Therefore, no recourse to analogy, a higher-level strategy, was necessary: cognitive systems are 'economically minded' and normally prefer to activate the lowest-level strategy possible for any given task. For students, however, with much less experience, even these well-defined problems were new and were therefore treated as ill-defined, at least to a certain extent. A low-level strategy such as contemplating visual displays was not sufficient and therefore when they could enlist analogy, which they did only when they were instructed to do so, it helped them solve the problem.

In contrast the ill-defined, or ill-structured problems were novel to all subjects and therefore analogy came in handy as a high-level design problem solving strategy. These results confirm our hypotheses regarding the role of analogy in the solving of ill-structured problems. But we still have to address those findings that pertain to the impact, or lack thereof, of visual displays on the effectiveness of problem solving in our experiments. The results we have obtained may seem confusing and inconsistent, but we think they are not. In fact, we believe they provide real insight into visual thinking and its relationship with analogical reasoning.

We postulate that the effect of visual displays on problem solving mechanisms is variable in strength and usability, in accordance with the mode of processing the displays undergo. For the sake of simplicity we will talk about two modes of processing, although there might be, and probably is, a more flexible continuum. At the lower-level mode, visual displays serve as pointers, icon-like memory-aids that facilitate retrieval of information from memory. When that information is stable and securely stored in memory retrieval is easy and this simple aid may be sufficient to promote successful integration of the retrieved information in the process of problem solving. When, however, knowledge is not available, or is not very securely stored and therefore difficult to access, the icon-like reminders are of no avail unless the system is induced to process them in the second, higher-level mode. This is where imagistic reasoning (as opposed to perception) comes into play and possibly interactive imagistic reasoning only (we have not checked possible correlation with factors such as sketching). We must stress that this mode of processing, which requires more cognitive resources, is reserved only for those cases where lower-level strategies are insufficiently effective. If correct, this theory explains the results we have obtained. In the case of ill-structured problems all our subjects had to resort to the higher-level processing mode of the visual displays and indeed, in order to exploit the visual images which they understood to harbor potentially useful information, or clues, for beneficial moves. In the case of well-structured problems the two groups of subjects were in very different positions as problem-solvers. The experienced designers required only low-level processing modes, and the mere provision of visual displays was sufficient. The novices (both beginning and advanced students) were unable to make use of this economical low-level processing mode, and therefore contemplating the displays had no beneficiary effect. Only when forced to go into higher-level processing modes as a result of the requirement to use analogy, could they take advantage of the availability of the displays.

5 In Conclusion

The implications for design education are clear to us. People can be trained to maximize the processing resources they are endowed with. Novices can be shown (e.g. through feedback) how the use of a certain cognitive strategy, analogic reasoning in this case, can be beneficial to them. At the same time repeated use can help stabilize domain-knowledge in their memory such that in subsequent cases it might require less effort, i.e. a lower-level processing mechanism, to retrieve and activate that knowledge [7]. Students of analogy have shown that the process of analogic mapping and transfer is a two-way operation: from a known example to abstraction, and from abstraction to a new candidate-example in the realm of the problem situation in hand (e.g. DeJong 89; Goldschmidt 95; above – sections 2 and 3). Novices who exercise analogical reasoning eventually appropriate the abstractions, implicitly or explicitly, thereby gaining in knowledge and in the ability to apply it, as claimed by Gentner and Medina (98). The use of analogy therefore brings similarity-based and rule-based reasoning very close together, and enables them to interact successfully in problem solving. In this interaction reasoning by similarity is typically helpful as a triggering device in the early phases of a search for ideas, whereas reasoning by rules follows suite in subsequent phases of development, refinement and evaluation. An example in the context of architecture is Le Corbusier's design for the roof of the Ronchamp chapel. According to the architect's own testimony (Broadbent 73) the shape of the roof was first conceived as a simile of a crab's shell which had been lying on his desk for quite some time. Le Corbusier, who made frequent use of analogical reasoning, transferred the structural principle of the double-membrane shell to the roof. Later, however, adjustments had to be made to ensure the proper functioning of the new concrete shell to its role as a roof, taking into account insulation, drainage and so on, as well as aesthetic and structural properties concerning the very large overhangs that lend the building its special silhouette.

In the various design domains, where typically ill-structured problems limit the serviceabilty of rules in the process of problem solving, reasoning by visual analogy is undoubtedly a very important strategy. For novice designers, it is also a most beneficial strategy for learning to design, and in particular for the acquisition of new concepts.

---

[7] The use of between-domain analogies is much harder, if not impossible to train for. Their spontaneous elicitation in problem solving appears to be all but reserved to extraordinarily creative thinkers. Chupin (98) reports a pedagogic experiment in which students of architecture are given between-domain analogies with which they are required to work. Unfortunately the report does not give any information on the effect of this experimental procedure on students' performance.

6 Acknowledgments

A first version of sections 2 and 3 of this paper was presented in the Sixth European Workshop on Imagery and Cognition in Oslo, in August 1997, under the title*: Dutch beer and the great wall of China: On visual analogy and mental imagery*. Many thanks to Nico van Scherpenzeel for his generous contribution of time and his enthusiastic cooperation in the author's efforts to understand his thinking. The experiments reported by Casakin and Goldschmidt were conducted as part of Hernan Casakin's Doctoral dissertation, which was supported by a stipend from the Technion. The writing of this paper was partially supported by a grant to the author from the Fund for the Promotion of Research at the Technion.

References

Ariety S, (1976), *Creativity: The magic synthesis*, Basic Books, New York, NY.

Beveridge M, and Parkins E, (1987), Visual representation in analogical problem solving, *Memory & Cognition,* 15(3), 230-237.

Broadbent G, (1973), *Design in architecture*, Wiley, London.

Casakin H, (1997), *The role of analogy and visual displays in architectural design*, Doctoral dissertation, Technion – Israel Institute of Technology, Haifa.

Casakin H, and Goldschmidt G, (1999), Expertise and the use of visual analogy: implications for design education, *Design Studies,* 20(2), 153-175.

Casakin H, and Goldschmidt G, (2000), Reasoning by visual analogy in design problem solving: the role of guidance, *Journal of Planning and Design: Environment & Planning B*, 27, 105-119.

Chase W G, and Simon H A, (1973), The mind's eye in chess, W G Chase Ed. *Visual information processing*, Academic Press, New York, NY.

Chupin J-P, (1998), The analogical phases of architectural design in studio teaching, *Proceedings of conference on research in design education*, NC State University at Raleigh, April 1998. Herberger Center for Design Excellence, College of Architecture and Environmental Design, Arizona State University, Arizona.

de Bono E, (1971), *The use of lateral thinking*, Penguin Books, Harmondsworth, UK.

DeJong G (1989), The role of explanation in analogy; or, the curse of an alluring name, S Vosniadou and A Ortony Eds. *Similarity and analogical reasoning*, Cambridge University Press, Cambridge, 346-365.

Dunbar K, (1995), How scientists really reason: Scientific reasoning in real-world laboratories, in R J Sternberg and J E Davidson Eds. *The nature of insight.* MIT Press, Cambridge, MA, 365-395.

Duncker K, (1945), On problem solving. *Psychological Monographs*, 58 (whole No. 270).

Finke R, (1990), *Creative imagery: Discoveries and inventions in visualization*, Lawrence Erlbaum and Associates, Hillsdale, NJ.

Fodor J A, (1975), *The language of thought*, Crowell, New York, NY.

Gentner D, (1983), Structure-mapping: A theoretical framework for analogy, *Cognitive Science*, *7*, 155-170.

Gentner D, and Medina J, (1998), Similarity and the development of rules, S A Sloman and L J Rips, (eds.), *Similarity and symbols in human thinking*, MIT Press, Cambridge, MA, 177-212 (reprinted from *Cognition*, 65(2-3), 1998).

Gick M L, and Holyoak K J, (1980), Analogical problem solving, *Cognitive Psychology*, *12*, 306-355.

Gick M L, and Holyoak K J, (1983), Schema induction and analogical transfer, *Cognitive Psychology,* 15, 1-28.

Goldschmidt G, (1991), The dialectics of sketching, *Creativity Research Journal 4* (2), 123-143.

Goldschmidt G, (1994), Visual analogy in design, R Trappl Ed. *Cybernetics and systems '94*, World Scientific, Singapore, 507-514.

Goldschmidt G, (1995), Visual displays for design: Imagery, analogy and databases of visual images, A Koutamanis, H Timmermans and I Vermeulen Eds. *Visual databases in architecture*, Avebury, Aldershot, UK.

Gregory S A, (1966), Design science, S A Gregory Ed. *The design method*, Butterworths, London, 323-330.

Gruber H E, (1995), Insight and affect in the history of science, R J Sternberg and J E Davidson Eds. *The nature of insight,* MIT Press, Cambridge, MA, 397-431.

Helstrup T, and Anderson R E, (1991), Imagery in mental construction and decomposition tasks, R H. Logie and M Denis Eds. *Mental images in human cognition*, North Holland, Amsterdam, 229-240.

Holyoak K J, and Thagard P, (1995), *Mental leaps: Analogy in creative thought,* MIT Press, Cambridge, MA.

Kaufmann G, (1980), *Imagery, language and cognition*, Universitetsforlaget, Bergen.

Keane M T, (1988), *Analogical problem solving*, Ellis Horwood Ltd. & John Wiley & Sons, Chichester.

Koestler A, (1964), *The act of creation*, Macmillan, New York, NY.

Kosslyn S M, (1994), *Image and brain*, MIT Press, Cambridge, MA.

Kosslyn S M, (1999), Visual mental images as re-representations of the world, in J S Jero and B Tversky Eds. *Visual and spatial reasoning in design*, Key Centre of Design Computing and Cognition, University of Sydney, Sydney, 83-92.

Langley P, and Jones R, (1988), A computational model of scientific insight, R J Sternberg Ed. *The nature of creativity*, Cambridge University Press, Cambridge, 177-201.

Martindale C, (1995), Creativity and connectionism, S M Smith, T B Ward and R A Finke Eds. *The creative cognition approach*, MIT Press, Cambridge, MA, 249-268.

Miller A I, (1986), *Imagery in scientific thought*, MIT Press, Cambridge, MA.

Perkins D N, (1981), *The mind's best work*, Harvard University Press, Cambridge, MA.

Peterson M A, (1993), The ambiguity of mental images: Insights regarding the structure of shape memory and its function in creativity, B Roskos-Ewoldsen, M J Intons-Peterson and R E Anderson Eds. *Imagery, creativity and discovery: A cognitive perspective*, North-Holland, Amsterdam, 151-185.

Rothenberg A, (1979), *The emerging Goddess*, University of Chicago Press, Chicago.

Schön D A, (1963), *The displacement of concepts*, Tavistock Publications, London.

Schunn C D, and Dunbar K, (1996), Priming, analogy, and awareness in complex reasoning, *Memory & Cognition,* 24(3), 271-284.

Shepard R N, (1978), Externalization of mental images and the act of creation, B S Randhawa and W E Coffman Eds. *Visual learning, thinking, and communication*, Academic Press, New York, NY, 133-189.

Simon H A, (1981, 2nd edition), The science of design, chapter 5, *The sciences of the artificial*, MIT Press, Cambridge, MA, 129-160.

Sloman S A, and Rips L J, Eds. (1998), *Similarity and symbols in human thinking*, MIT Press, Cambridge, MA (reprinted from *Cognition*, 65(2-3), 1998).

Sloman, S A, (1996), The empirical case for two systems of reasoning, *Psychological Bulletin*, 119(1), 3-22.

Verstijnen I M, van Leeuwen C, Goldschmidt G, Hamel R, and Hennessey J M, (1998), Creative discovery in imagery and perception: Combining is relatively easy, restructuring takes a sketch, *Acta Psychologica*, 99, 177-200.

Vosniadou S, and Ortony A, Eds. (1989), *Similarity and analogical reasoning*, Cambridge University Press, Cambridge, MA.

Weisberg R W, (1986), *Creativity: Genius and other myths*, W.H. Freeman & Company, New York, NY.

Weisberg R W, (1988), Problem solving and creativity, R J Sternberg Ed. *The nature of creativity*, Cambridge University Press, Cambridge, 148-176.

Wotiz J H, and Rudofsky S, (1984), Kekulé's dreams: Fact or fiction? *Chemistry in Britain*, 720-723.

Wotiz J H, ed. (1993), *The Kekulé riddle: A challenge for chemistry and psychology*, Cache River Press, Boulder, CO.

Chapter 10

Dissociation of Design Knowledge

Vinod Goel
Dept. of Psychology
York University
Toronto, Canada

Abstract

It is often noted that successful designing requires two types of knowledge: explicit, articulate, domain-specific knowledge, and inarticulate, domain-independent, procedural knowledge. Neuropsychological evidence is provided for a dissociation between the two types of knowledge. This reinforces the reality of the distinction. It is further suggested that the inarticulate, procedural knowledge is not knowledge in the traditional sense of the word, but consists of *mechanisms* to support "ill-structured" representations and computations. This implies that learning environments – like the design studio – that provide students with *practice* in groping and coping with ill-structured, real-world design problems must be an integral part of design education.

> *Let him be educated, skillful with pencil, instructed in geometry, know much history, have followed the philosophers with attention, understand music, have some knowledge of medicine, know the opinions of the jurists, and be acquainted with astronomy and the theory of the heavens.*
> --Vitruvius (on the education of the architect), quoted in Broadbent (1973, p. 4)

1 Introduction

The organizers of the conference that resulted in this book proposed that there are two types of design knowledge:

> One kind of knowledge [Type 1] is domain-related and technical; how does a building stand up? What are the basic components of a compiler or graphics system? What are the procedures in a hospital? How are computer chips fabricated? Another kind of knowledge [Type 2] seems less domain-specific and seems largely procedural. It deals *with managing the processes of designing: how*

> *to structure and define a problem, how to identify new issues for an already defined problem? how to generate non-obvious alternatives?* The second kind of knowledge cannot be easily measured, and the effectiveness of the knowledge *interacts with other dimensions of a person, such as personality, social skills and self-discipline.* [italics added]

I am not sure that the declarative / procedural categories are rich enough to capture the distinction, but I am sure that there is a crucial distinction to be captured. Type 1 knowledge is easily identified, organized, articulated and disseminated in textbooks and classroom lectures. Type 2 knowledge is more illusive. It is passed down in more subtle, inarticulate ways.

The main goal of this chapter is to argue that the distinction between Type 1 and Type 2 knowledge has a neuropsychological basis, as demonstrated by anatomical dissociation, and to show the consequences on the design process of a deficit in Type 2 knowledge. We begin by discussing the importance of finding a neuropsychological dissociation corresponding to an intuitive or psychological distinction; review some famous cases of patients that seem to exhibit a dissociation between Type 1 and Type 2 knowledge; review the performance of an architect with a lesion to the right prefrontal cortex, that seems to leave Type 1 knowledge intact but grossly impairs Type 2 knowledge, rendering him incapable of designing; explain the deficit in terms of theoretical ideas developed in Goel (1995); and conclude by drawing some lessons for design education.

2 Importance of Neuropsychological Dissociations

The proposed distinction between Type 1 and Type 2 design knowledge is an intuitive, functional distinction, perhaps based on some psychological theory. But we know that our intuitions are often wrong and that psychological theories are notoriously underconstrained. More generally, functional individuations are not causally constrained. We usually use such individuations just when we do not know the underlying causal structure. This devalues the currency of functional distinctions. But if we can show that our functional distinctions map onto causally individuated neurophysiological structures, then we can have much greater confidence in the functional individuation.

By way of an example, suppose that we individuate the following three functions on the basis of behavioral data: (f1) raise left arm, (f2) raise left foot, (f3) wiggle right ear. If these functions can be mapped onto three causally differentiated structures in a one-to-one fashion, we would be justified in claiming to have discovered three distinct functions. If, however, all three of our behaviorally individuated functions map onto one causally differentiated structure, in a many-to-one fashion, we would say that our functional individuation was too fine grained and collapse the distinctions until we achieved a one-to-one mapping. That is, raising the left arm does not constitute a distinct function, but the conjunction of the three do constitute a single function. If we encountered the reverse situation, where one behavioral function mapped onto several causally distinct structures, we would conclude that our individuation was too coarse-grained and refine it until we achieved a one-to-one mapping. One final possibility is a many-to-many mapping between our functional individuation and casually individuated structures. Here we would have a total cross-classification and would have to assume that our functional individuations are simply wrong and start over again.

One way of discovering these mappings is through studies of patients with brain lesions. Brain lesions result in selective impairment of behavior (cognitive and otherwise). Such selective impairments are called dissociations. A single dissociation occurs when we find a case of a lesion in region *x* resulting in a deficit of function *a* but not function *b*. If we find another case, in which a lesion in region *y* results in a deficit in function *b* but not in function *a*., then we have a double dissociation.

The most famous example of a double dissociation comes from the domain of language. In the 1860's Paul Broca described patients with lesions to the left posterior inferior frontal lobe who had difficulties in the production of speech but were quite capable of speech comprehension. This is a case of a single dissociation. In the 1870's Carl Wernicke described two patients (with lesions to the posterior regions of the superior temporal gyrus) who had difficulty in speech comprehension, but were quite fluent in speech production. The addition of this observation results in a double dissociation.

With respect to design knowledge, if we are presented with a patient in whom a lesion in area *x* impairs the ability to access Type 2 design knowledge, but allows unimpaired access to Type 1 design knowledge, then we have a case of single dissociation. If other cases are found where a lesion in area y impairs the ability of the patient to access Type 1 knowledge, but allows unimpaired access to Type 2 design knowledge, then we have evidence for a double dissociation. In either case we have evidence for differences in causal mechanisms underlying Type 1 and Type 2 knowledge. In the case of a double dissociation we would have evidence for two distinct neurophysiological mechanisms.

3 Relevant Cases

The most relevant cases for us come from the frontal lobe patient literature. There are many reports of patients, who are by in large neuropsychologically intact, but cannot successfully function in the world. The literature originates with Harlow's (1868) classic description of Phinas Gage. Gage was a railroad foreman who had a four foot iron tamping rod pass through his skull in a freak accident. Gage survived, and in many respects continued with his life. However, Harlow noted that prior to the injury Gage was mild mannered and "considered a shrewd businessman, energetic & persistent in executing all his plans." But after the injury he would devise "many plans of future operation but discarding them before they were executed." He was also fitful, irreverent, obstinate, and impatient of restraint or advice. There seemed to be an imbalance between his "intellectual faculty" and "animal propensities". His friends said of him that he was "no longer Gage."

Another famous description is Penfield's (Penfield and Evans 35) report on the difficulties in planning dinner experienced by his sister fifteen months after the removal of her right frontal lobe:

> When the appointed hour arrived she was in the kitchen, the food was all there, one or two things were on the stove, but the salad was not ready, the meat had not been started and she was distressed and confused by her long continued effort alone. It seemed evident that she would never be able to get everything ready at once... Although physical examination was negative and there was no change in personality or capacity for insight, nevertheless the loss of the right frontal lobe had resulted in an important defect. The defect produced was a lack of capacity for planned administration.... (p. 131)

In the more recent literature, Eslinger and Damasio (85) describe the patient EVR, a successful 35 year old accountant who underwent an operation for the removal of a large orbitofrontal meningioma (a type of tumor originating in the meninges, the protective membrane surrounding the brain). After the operation EVR tested in the above average range to superior range on IQ and memory tests, yet was unable to hold a job and function successfully in the world (p. 1731-1732):

> After a 3-month recovery period, he returned to accounting and bookkeeping.... He soon became involved in a home-building partnership with a former coworker, a man of questionable reputation who had been fired from the company... The business failed and he had to declare bankruptcy..... Thereafter, he drifted through several jobs. He worked as a warehouse laborer, as a building manager, and as an accountant ... but was fired from each. Employers complained about tardiness and disorganization, although basic skills, manners, and temper were appropriate. Similar difficulties led to a deterioration of his marital life.... Unable to hold a job and separated from his family, EVR moved in with his parents....Employment problems continued.... He needed 2 hours to get ready for work in the morning, and some days were consumed entirely by shaving and hair-washing. Deciding where to dine might take hours... He would drive to each restaurant to see how busy it was.... Purchasing small items required in-depth consideration of brands, prices, and the best method of payment. He clung to outdated and useless possessions, refusing to part with dead houseplants, old phone books, six broken fans....

All of these observations speak of difficulties in judgment, decision-making, and problem solving in real-world, open-ended situations. Often the problem solving involves planning and look-ahead components. Perhaps the most relevant case for our purposes is that of patient PF, an accomplished architect, reported by Goel and Grafman (00). It provides an illustration of what happens to the design process when Type 2 design knowledge is unavailable.

4 Patient PF: Impairment of Type 1 Knowledge[16]

Goel and Grafman (00) report the case of patient PF, a right-handed 57-year-old Caucasian male architect, who suffered from a predominantly right hemisphere lesion to the prefrontal cortex (though some minor anterior left frontal damage was also present).

PF earned a graduate degree in Architecture from Yale University. He scored in the 98th percentile on the Graduate Record Examination in math and science. After graduation he practiced in the USA until the mid-1970s and then moved to Spain to work on a long-term "dream project". He was successfully self-employed in Spain until his illness. At the time of testing, PF had IQ and memory scores in the superior to excellent range, his ability to draw was intact, yet he was unable to design. He was involuntarily retired and lived at home with his mother.

To explore the nature of his deficit, Goel and Grafman matched him for age and education with a 54 year old Architect and engaged both in a simple architectural design/planning task. Both subjects were asked to propose a redesign for our lab space. They were given the following problem statement specifying a set of constraints and encouraged to ask for additional information as necessary:

> Our lab space is located in Room 5D51. It currently houses three scientists and five research assistants. Another scientist is expected in January. The number of research assistants can increase up to 16 during the summer months.
> The space is used for reading, writing, computing, telephone conversations, and so on. In fact, we do all of our work in this space except for seeing patients. Some of us spend up to 10 hours per day there. It is a very dismal environment.
> Your task is to reorganize, redesign, reconfirm the space such as to increase our comfort and productivity. We do not have a budget for the redesign. However, we do have the option of exchanging some of our furniture at the surplus store, and perhaps we can pool personal time and resources to do some painting and cleaning.
> You have two hours to propose a design. You may spend up to 15 minutes of the first hour in the lab space. While there, you may measure, make notes and sketches, and ask anyone there any questions you think relevant. You may revisit the lab for 10 minutes anytime during the second hour. Please begin.

[16] Much of this section (including Figures 1 – 4) is reproduced from Goel and Grafman (2000).

Both subjects reported having experience designing office spaces and noted that this was a very easy problem for them.

Table 1: Itemization of modules and submodules utilized by patient (PF) and control (NC).

PF	NC		PF	NC	
x	x	Users	x	x	Communal Facilities & Space
x	x	numbers	x	x	furniture
x	x	class/ categories	x	x	filing cabinets
x	x	relationship between users	x	x	misc equipment
x	x	changes over course of year	x	x	printer
x	x	work habits		x	message board
x		source of discomfort		x	photocopying
x	x	Activities in space	x	x	closet space
x	x	experiment design & analysis	x	x	storage space
x	x	reading & writing	x	x	entrance zone
x	x	meetings	x	x	dimensions
x	x	Social Concerns	x	x	configuration
x	x	social clustering/ social hierarchy	x	x	refrigerator
x	x	community space	x	x	microwave
x	x	private space		x	Lighting
x	x	character of space		x	natural light
	x	personalization of space		x	task lights/ personal lamps
x	x	noise levels		x	florescent lights
x	x	Circulation Patterns	x	x	Border Conditions
	x	cul de sac/	x	x	walls
	x	dead space		x	ceilings
	x	congestion		x	windows
x	x	Private Facilities & Space	x	x	door
x	x	workstations		x	flooring
x	x	computers		x	venting
	x	seating		x	orientation (N, S, E, W)
x	x	dimensions	x	x	dimensions
x	x	configuration	x		Cleaning/maintenance

The first thing to note about PF's performance is that his explicit (Type 1) knowledge of design is intact. An episode level analysis of the data provides a measure of goals/subgoals and strategies pursued by subjects. It is a content analysis. It gives an indication of the knowledge the subjects bring to the problem space and the issues they consider. Table 1 shows the problem solving episodes engaged in by our patient and control subject. As can be seen in Table 1, both subjects considered issues ranging from information about users (e.g. numbers, categories), their goals (e.g. a quieter environment), behaviors/activities the users need to engage in (e.g. meetings, writing), the functions the artifact/space needs to support (e.g. circulation patterns), and the actual structure of the artifact (e.g. printers, workstations, dimensions). Both the control and patient displayed the sophisticated knowledge base one would expect from experienced architects.

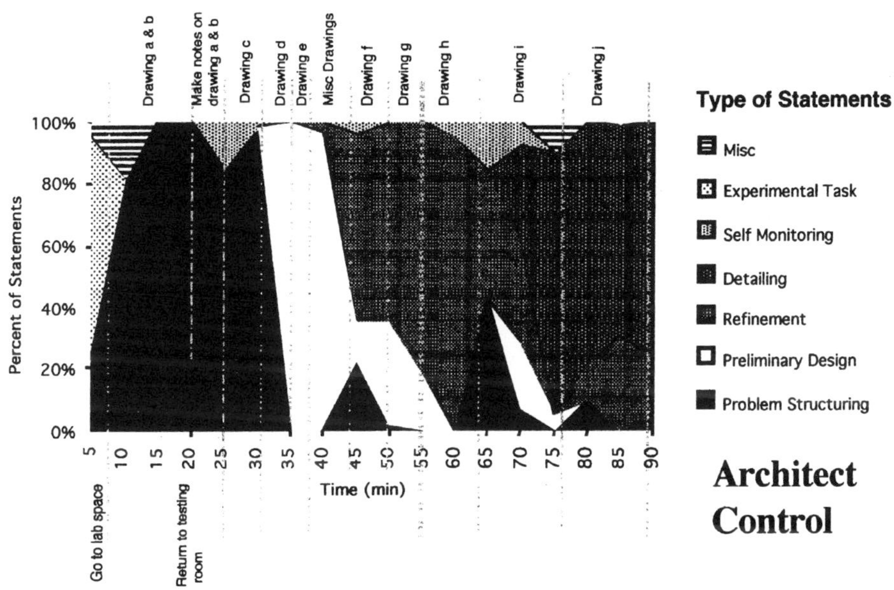

Figure 1: Control's design-development statements aggregated into 5 minute intervals. Drawings and key events are overlaid on top.

However, there are significant differences in the problem solving behavior of the patient and control. Figures 1 and 2 show the temporal distribution of design-level statements for the control and patient, aggregated over five-minute intervals. The control's temporal distribution of statements is quite typical of designers (Goel 95). He begins by problem structuring and has it largely completed in the first quarter of the problem solving session.

The next part of the session is devoted to preliminary design, followed by design refinement and then detailing, each in roughly equal proportion. The whole process is accompanied by some self-monitoring and the phases reoccur as needed throughout the session.

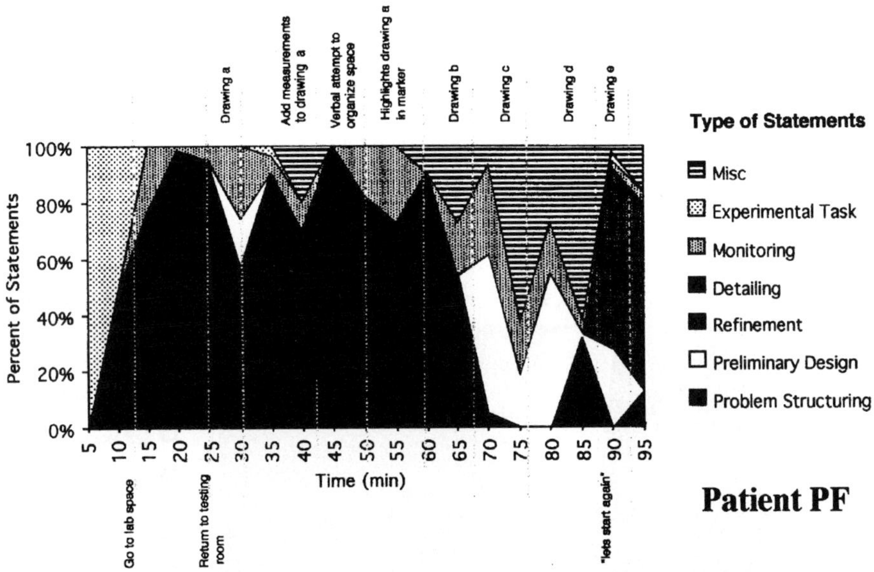

Figure 2: Patient's design-development statements aggregated into 5 minute intervals. Drawings and key events are overlaid on top.

The patient's progress through the problem space is quite different from that of the control subject (Figure 2). The patient spends most (two-thirds) of the session in problem structuring mode. The preliminary design phase, occurs at the tail end of the session. Its duration is short; it is accompanied by much self monitoring and a great many miscellaneous statements; and as we will see shortly, the patient's movement through this phase is erratic and lacks progression. It is a phase that the patient has enormous difficulty with. At the 87 minute mark, the beginning of the refinement period, the patient notes "lets start again." Refinement of this proposal occurs during the last few minutes and the session ends abruptly.

The 'x' indicate utilizationOverlaid on top of the temporal distribution of design-level statements in Figures 1 and 2 are the key drawings made the by the subjects and some highlights of the session. Drawings depict the generation and development of ideas/concepts. As such they constitute states in the designers' problem space. Figures 3 and 4 show the actual drawings made by the subjects and the type of transformations that lead from one drawing/idea to the other.

There are nine states in the control's problem space (Figure 3), a start state (a & b), a goal state (j) and seven intermediate states (c - i). The starting state for the control is drawing (a) of the existing lab space and the accompanying measurement drawing (b). Both drawings were made in the lab space. Three of the intermediate states belong to the preliminary design phase (c - e); four belong to the refinement phase (f - i); and the final state belongs to detailing (j). The preliminary design states are all quite abstract. He begins by considering "circulation patterns" in drawing (c). This pattern constitutes his kernel idea. It is developed and transformed to deal with the issue of and "social organization" (d) and "permanent and transient spaces" (e). The refinement drawings are structural. They depict workstations, tables, doors, windows, and corridors. The subject transforms state (e) into a proposal (drawing f) (half way during the session) that he considers "reasonable". However he thinks the center condition can be improved. He therefore holds the perimeter conditions constant and transforms the center in drawing h. He rejects drawing (h), returns to drawing (f) and transforms it into drawing (g). He is happy with the idea depicted by drawing (g). He now shifts gear and begins to detail and fine tune the proposal, first in section (drawing i) and then in plan (drawing j).

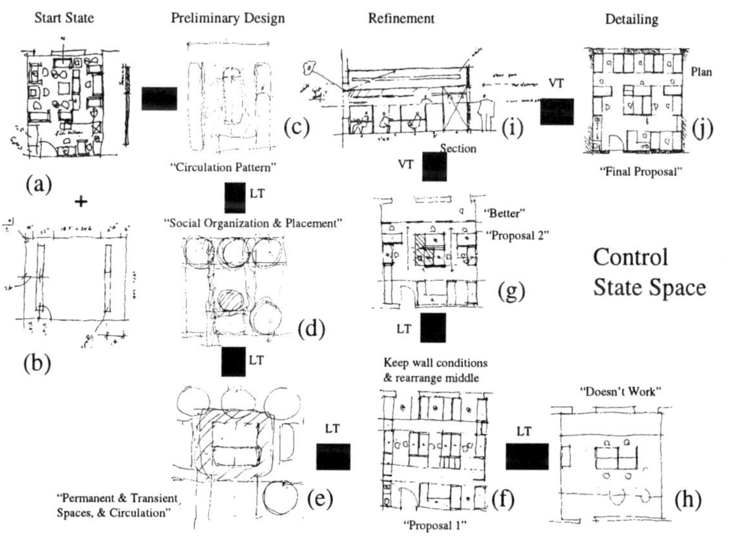

Figure 3: Control's state space and transformation functions. LT = lateral transformation; VT = vertical transformation.

The movement from states (c) to (g) is underwritten by lateral transformations. A *lateral transformation* is one where movement is from one idea to a slightly different idea rather than a more detailed version of the same idea (Goel 95). The transformation of state (g) to (i) and from (i) to (j) is underwritten by vertical transformations. A *vertical transformation* is one where movement is from one idea to a more detailed version of the same idea (Goel 95).

An analysis of the patient's state space tells a very different story. There are five states in the patient's problem space (Figure 4), a start state (a), a goal state (e) and three intermediate states (b - d). The start state drawing of the existing lab space (a) was completed by the patient from memory in the testing room. It is as detailed and accurate as the control's drawing. The patients final state drawing (e) was completed during the refinement phase. The three intermediate drawings (b - d) were completed during the preliminary design phase. The first of these drawings (c) – the kernel idea – occurs two-thirds way into the session. Unlike the preliminary drawings of the control, the patient is concerned with arranging furniture right from the start. But perhaps the most dramatic difference between the patient and control is that the patient's three preliminary design drawings are fragmentary and unrelated.

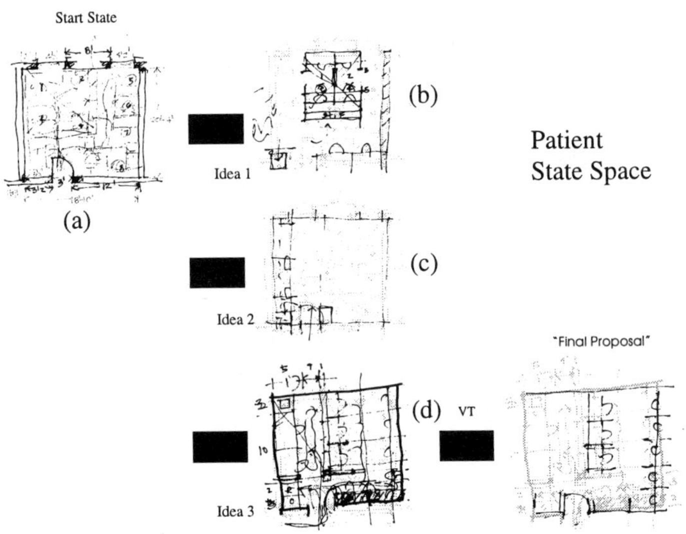

Figure 4: Patient's state space and transformation functions. LT = lateral transformation; VT = vertical transformation. Drawings (b) to (e) were executed on transparent tracing paper on top of drawing (a).

Preliminary design sketches are, almost by definition, fragments of ideas. Designers do not generate several independent fragments and choose between them. They generate a single idea/fragment and develop it through transformations (lateral or vertical) to a point it is complete and can be evaluated (Goel 95). The patient has made several (successful) attempts to generate idea fragments. But he is unable to develop and explore these ideas through the application of lateral transformations. Each of his preliminary drawings must be treated as independent idea/fragments. Indeed, he tries to articulate the difficulty he is experiencing as follows:

> You see, normally, what I would have, even as a student, I'd be -- there would be sketches on top of sketches. And I could -- it would be progressive. Here I seem to be doing several different thoughts on the same piece of paper in the same place, and it's confusing me. So, instead of the one direction that I had at the beginning, I have three or four contradictory directions with not a kind of anchor to work from....

On generating drawing (d) 87 minutes into the session, he says "lets start again". The only idea/drawing transformation he engages in is from (d) to (e). This is a vertical transformation where he fine-tunes the proposal.

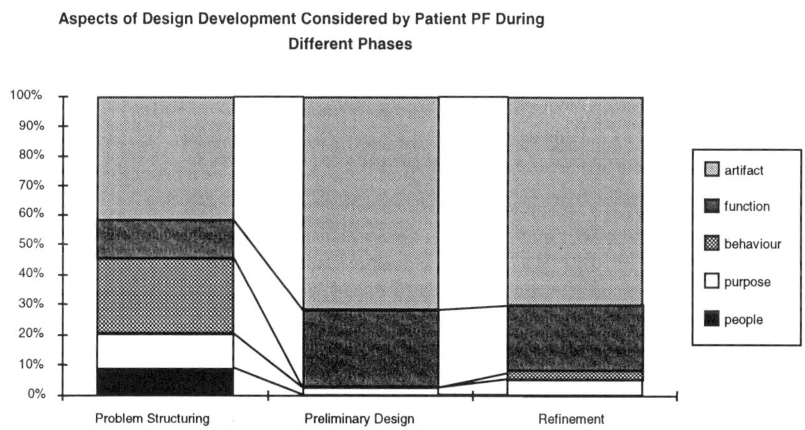

Figure 5: Aspects of design development considered by control subject during different phases of problem solving

Further evidence of the patient's difficulties with the preliminary design phase is provided by analyzing the "aspects of design development" considered by the control and patient during the four phases of design development. Figure 5 shows the percentage of design development statements (problem structuring, preliminary design, refinement, and detailing) the control subject devotes to the consideration of people (users), their goals/purposes, the behaviors that need to be supported, the function the artifact needs to serve, and the structure of the artifact itself. One can see that during problem structuring and preliminary design the majority of the statements are devoted to abstract, nonstructural aspects of the design having to do with users, goals, behaviors, and functions. This distribution shifts with the refinement phase, where function and structure dominate, and again in the detailing phase, where structure dominates. For example, during the structuring phase the control considers issues of social hierarchy (research assistants vs. scientists). This is propagated and translated into structural arrangement in terms of giving the scientists more spacious workstations and placing them in corner locations.

Figure 6 shows the corresponding data for patient PF. It contains only three phases (because the patient did not enter into a detailing phase). The level of abstractness of the statements in the patient's problem structuring phase is quite similar to the control's. He solicits and generates much relevant information about people, purpose, and behaviors. The patient's refinement phase distribution is also not unlike the control's. However, the patient's preliminary design phase is dramatically different. Whereas the control carries through the information regarding people, purpose and behavior through to the preliminary design phase (and in fact through to refinement and detailing) and uses it to guide the emerging design, *the patient is unable to carry through this information and use it to guide problem solving.*

PF provides an intriguing example of a trained architect whose explicit knowledge of design is intact (as evidenced by the episode-level analysis in Table 1) but he is unable to participate in even very simple design problem solving activities. I think that the breakdown we are observing is quite consistent with the Type 1 and Type 2 distinction in design knowledge that we began with. His Type 1 knowledge is intact but he is unable to access Type 2 knowledge. I would now like to review four neuropsychological theories that attempt to explain the observed deficit.

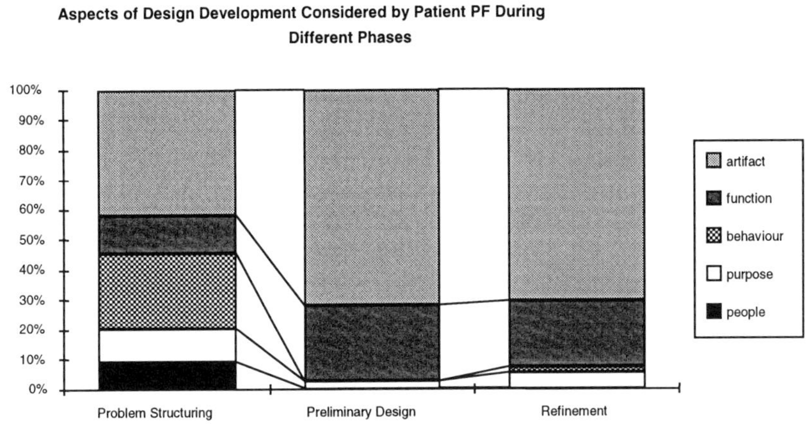

Figure 6: Aspects of design development considered by patient during different phases of problem solving

5 Theories of Frontal Lobe Function

There are several theories of frontal lobe dysfunction in the current literature that try to explain the types of deficits encountered by PF, including somatic markers (Damasio, 1994), structured-event complexes (Grafman 89), and supervisory attentional system (Shallice 88).

Damasio (94) argues that the major issue with patients with medial ventral frontal lobe lesions is that they make poor judgments in real-world situations. He also notes that such patients have severe emotional and social problems. They seem detached, unconcerned, and uninvolved. Damasio makes a causal connection between emotional deficits and poor judgment. His intuition is that there must be a close integration of the cognitive and the emotional. He suggests that the cause of patient's poor judgments is that these patients are unable to *inform* cognitive/reasoning processes by visceral, noncognitive, emotional factors, which he calls "somatic markers".

> In short, *somatic markers are a special instance of feelings generated from secondary emotions.* Those emotions and feelings *have been connected, by learning, to predicted future outcomes of certain scenarios.* When a negative

> somatic marker is juxtaposed to a particular future outcome the combination functions as an alarm bell. When a positive somatic marker is juxtaposed instead, it becomes a beacon of incentive. (Damasio 94, p. 174)

The structured-event complex (SEC) theory (Grafman 89) proposes that much of our world knowledge is stored in script- or episode-like data structures and frontal lobe patients have difficulty in retrieving these structures. Scripts are large-scale knowledge structures that are thought to account for much of our routine behavior. Formally the representations – called managerial knowledge units (MKUs) – are hierarchically ordered, linked list data structures. The theoretical properties include varying thresholds of operations, temporal properties, categorical distinctions, multiple formats of representation, combinatorial binding with other forms of knowledge, and a relational cognitive architecture. The hypothesis is that frontal lobe dysfunction results in an impairment in the ability to access and retrieve these data structures. Its basic prediction is that frontal lobe patients will have more difficulty with knowledge-rich tasks than knowledge-impoverished tasks.

The supervisory attentional system (SAS) theory (Shallice 88) tries to account for patients' differential performance in routine and novel tasks by postulating dual control mechanisms. The idea is that there is a built-in contention scheduler that determines responses in over-learned, routine situations. However, when the organism is confronted with a novel situation, the contention scheduler is unable to cope and passes control to the more sophisticated supervisory attentional mechanism. Shallice's claim is that the frontal lobes serve as this meta-level control mechanism.

While there is much to be said for these theories, the problem for our current purposes is that they do not give us any reasons to believe that these mechanisms are differently implicated in the early and later phases of design problem solving. That is, nothing in these theories predicts that patients should have more difficulty in the preliminary design phase than the later more structured phases.

Goel (95) characterizes design problem solving as involving four development phases: problem structuring, preliminary design, refinement, and detail specification; and notes that each phase differs with respect to the type of information dealt with, the degree of commitment to generated ideas, the level of detail attended to, the number and types of transformations engaged in, and the symbol systems needed to support the different types of information and transformations. This is depicted in Figure 7.

What is of interest to us here is the contrast between preliminary design and refinement and detailing phases. Preliminary design is a classical case of creative, ill-structured problem solving. It is a phase where alternatives are generated and explored. This generation and exploration of alternatives is facilitated by the abstract nature of information being considered, a low degree of commitment to generated ideas, the coarseness of detail, and a large number of lateral transformations. A lateral transformation is one where movement is from one idea to a slightly different idea rather than a more detailed version of the same idea. Lateral transformations are necessary for the *widening of the problem space* and the exploration and development of kernel ideas (Goel 95).

The refinement and detailing phases are more constrained and structured. They are phases where commitments are made to a particular solution and propagated through the problem space. They are characterized by the concrete nature of information being considered, a high degree of commitment to generated ideas, attention to detail, and a large number of vertical transformations. A vertical transformation is one where movement is from one idea to a more detailed version of the same idea. It results in a *deepening of the problem space* (Goel 95).

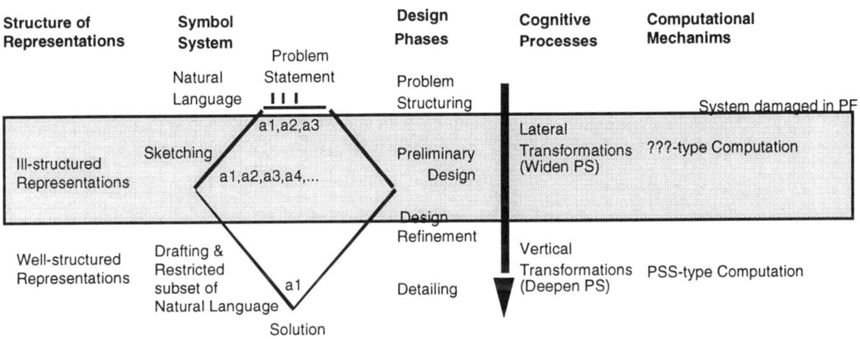

Figure 7: The design problem space

Figure 7 also notes that preliminary design phases are accompanied by "ill-structured" representations (in this case belonging to the symbol system of sketching), while the refinement and detail phases are accompanied by more "well-structured" representations, belonging to the system of drafting (Goel 95). Intuitively one might understand "well-structured" mental states as being precise, distinct, determinate, and unambiguous. "Ill-structured" mental states, on the other hand are imprecise, ambiguous, fluid, amorphous, indeterminate, vague, etc. A formal treatment appears in Goel (95).

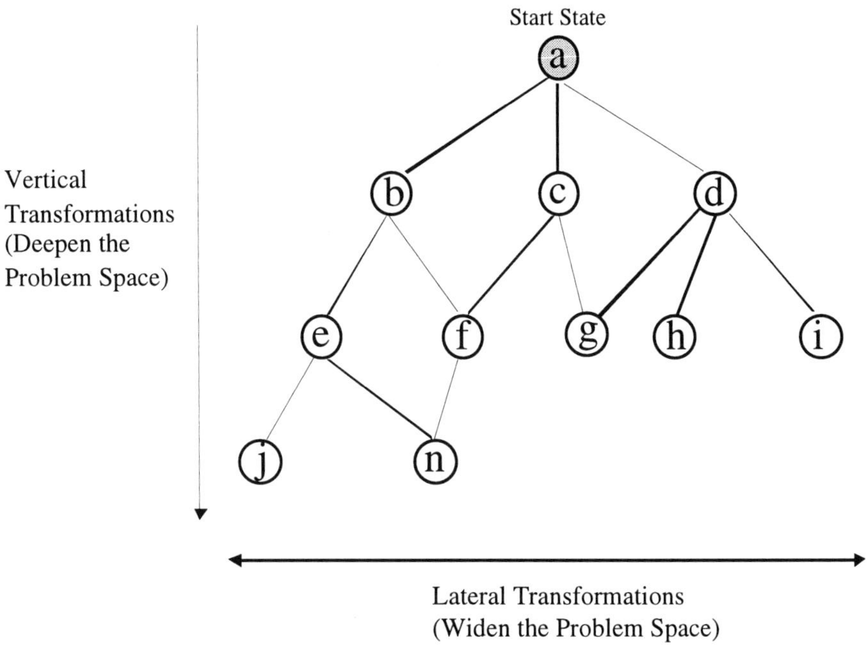

Figure 8: Structure of state space deemed to be necessary for well-structured problems

Some of these properties of representations are illustrated informally in Figures 8 and 9. Figure 8 shows a well-structured state space and how it is developed in by vertical and lateral transformations. I would like to pick out three properties of this problem space: (i) no state overlaps with any other state; (ii) there is a measurable amount of "distance" between states; (iii) the referent of a state does not change with context (i.e. with where it appears in the graph) so the states are unambiguous.

In contrast to the state space in Figure 8, the state space in Figure 9 is ill-structured. (i) There are overlapping states such as *g* and *h*. (ii) There are "densely" ordered states – with little or no measurable "distance" between them – such as *j* through *m*. (iii) States in such problem spaces can have different referents depending on their location in the graph and are therefore ambiguous.

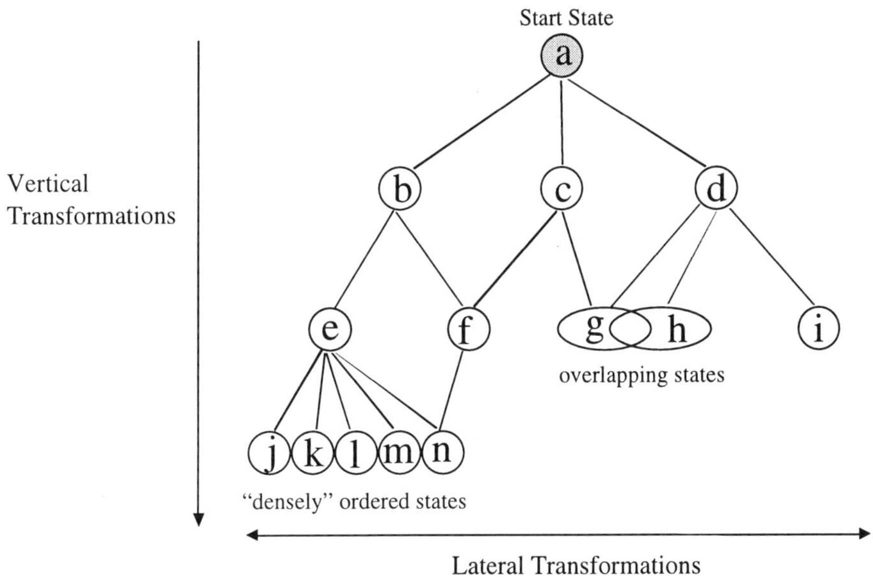

Figure 9: Structure of state space deemed to be necessary for ill-structured problems

There is empirical evidence that these properties of ill-structured representations facilitate lateral transformations in the following ways (Goel 95):

1) The overlapping of states introduces a degree of coarseness into the problem space by allowing a single state to simultaneously instantiate a number of ideas in a noncomplex or primitive manner. The presence of overlapping states allows the problem solver to remain non-committal about the state he/she is in.

2) The dense ordering of states gives the problem space degree of fine-grainedness. This reduction in "distance" between states allows for the coverage of the full range of possibilities and facilitates the transformation from one state to another.

3) The ambiguity of states insures that the referents of the states during the early phases of design are indeterminate. Ambiguity is important because one does not want to crystallize ideas too early and freeze design development. This is graphically summarized in Figure 10.

I am arguing that the crucial cognitive/computational differences between the phases of planning are to be found in the structure of the representations and computations associated with different planning phases. Well- and ill-structured representational system are needed to underwrite well- and ill-structured planning phases respectively. It can further be demonstrated that ill- and well-structured representations support different types of computational mechanisms (Goel 95). Well-structured representations are necessary for the "classical" von Neumann-type computational devices postulated by cognitive science. I have referred to these as PSS-type (Physical Symbol System) computation in Figure 7. Ill-structured representations support computations in different types of systems, perhaps multi-layered connectionist networks. These are referred to as ???-type computation in Figure 7.

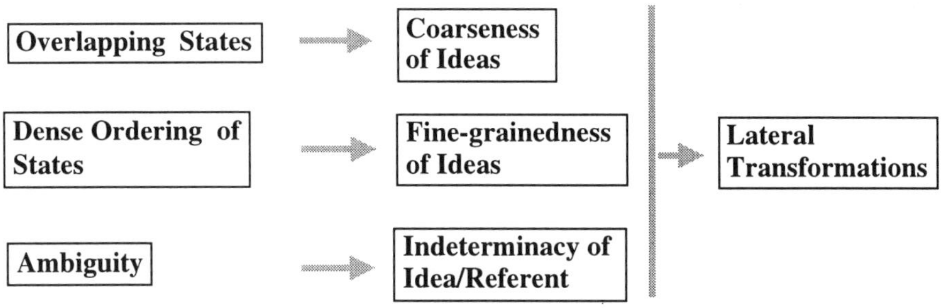

Figure 10: Role of ill-structured mental states in lateral transformations

Given this theoretical apparatus and the behavioral deficits observed in the patient, we postulate that PF's lesion has resulted in a selective impairment of the neural system that supports "ill-structured" representations and computations (see Figure 7). The theory allows that the "well-structured" representation and computation system may be largely intact and predicts that the patient may be competent at the detailing (well-structured) phase of problem solving where such a system is required. But because his preliminary design phase is unsuccessful, it is hard to make much of later phases. However, his exceptional performance during problem structuring and on the well-structured neuropsychological tests suggests that the computational system used for well-structured tasks may be intact.

The claim about ill- and well-structured representational and computational systems is a general claim about the structural and computational properties of mental states (Goel 95). The formal properties differentiating ill-structured and well-structured representations do not break down along linguistic and pictorial lines. So there is no reason to expect that the explanation offered is specific to design or drawing domains, though the system is particularly well developed in designers and they can provide an external trace of it. One would expect to find ill-structured representations and computations used in many ill-structured problem solving domains.

6 Conclusion

I have provided data to suggest that the intuitive/behavioral distinction between Type 1 and Type 2 design knowledge runs deep. It has a neuroanatomical basis. The data suggest that while the explicit Type 1 knowledge may be necessary for successful designing, it is clearly not sufficient. When Type 2 knowledge is absent, patients literally cannot design. Furthermore, I have argued that this "knowledge" is not knowledge in the conventional sense. On my account, Type 2 knowledge consists in having an intact, well developed neural mechanism to deal with "ill-structured" situations. As a mechanism, it is not something to be learned and remembered, like a formula for calculating a point load on a beam. Rather, given the neural endowment, the educational challenge is to develop and hone it through practice. On this account it is not surprising that Type 2 knowledge should seem inarticulate and elusive.

This view has consequences for design education. "Teaching" students Type 2 design knowledge is a matter of giving them the opportunity and encouragement to develop the representational and computational mechanisms necessary to deal with ill-structured situations. The data and arguments I have presented do not speak to *how* such opportunities should be provided. My own personal view is that the best strategy may be to put students in situations that *simulate* the ill-structured, incomplete, and under-constrained nature of real-world design problems and allow them to *grope* their way through (with some minimal guidance). There is, of course, nothing new in this proposal. It just reinforces the importance of traditional design studios. The reason they work so well – and must remain an integral part of design education – is that they provide students with practice in the deployment and development of the relevant neural mechanisms necessary to cope with ill-structured situations.

References

Bechara A, Damasio A R, Damasio H, and Anderson S W, (1994), Insensitivity to future consequences following damage to human prefrontal cortex, Cognition, 50, 7-15.

Broadbent G, (1973), *Design in architecture: architecture and the human sciences*, John Wiley & Sons, New York, NY.

Damasio A R, (1994), *Descartes' Error*, Avon Books, New York, NY.

Eslinger P J, and Damasio A R, (1985), Severe disturbance of higher cognition after frontal lobe ablation: Patient EVR, *Neurology*, 35, 1731-1741.

Goel V, (1995), *Sketches of Thought*, MIT Press, Cambridge, MA.

Goel V, and Grafman J, (2000), The role of the right prefrontal cortex in ill-structured problem solving, *Cognitive Neuropsychology*, 17(5), 415-436.

Grafman J, (1989), Plans, actions, and mental sets: Managerial knowledge units in the frontal lobes, E. Perecman Ed. *Integrating Theory and Practice in Clinical Neuropsychology*, Lawrence Erlbaum and Associates, Hillsdale, NJ.

Harlow J M, (1868), Recovery after severe injury to the head, *Publications of the Massachusetts Medical Society*, 2, 327-346.

Penfield W, and Evans J, (1935), The frontal lobe in man: A clinical study of maximum removals, *Brain*, 58, 115-133.

Shallice T, (1988), From neuropsychology to mental structure, Cambridge University Press, Cambridge, MA.

Chapter 11
Scenario-based design: A brief history and rationale

John M. Carroll
Department of Computer Science and
Center for Human-Computer Interaction
Virginia Tech
Blacksburg, Virginia

Abstract

Envisioning and analyzing scenarios of human activity can help us to understand and to create appropriate artifacts to support human activity. Scenario-based design offers significant and unique leverage on some of the most characteristic and vexing challenges of design work. By emphasizing reflection and inquiry as design activities, and by helping to integrate diverse design activities, scenario-based design may also be particularly suitable to motivate and guide design learning.

1 Scenarios in the design of computer systems and software

In the traditional design of computer software and systems there is a key document called a functional specification. The functional specification enumerates the functional components of the system, and the components of the components. This decomposition terminates with lists of functions and quite detailed descriptions of how each function is invoked and what it does.

An interesting thing happened to functional specification documents during the 1980s. First, scenarios of use started appearing in the appendices of these documents. These scenarios were usually textual narratives describing a typical interaction with the system specified in the main body of the document. They were stories of use: A manager is organizing her schedule for the week; she opens the Personal Planner tool, opening Monday in one window and her high-priority to-do items in another window; and so on. Sometimes these scenarios were presented as storyboards, that is, as a series of sketches depicting workplace scenes and computer display states with associated textual descriptions.

Through the decade of the 1980s, these scenarios began a march forward through the functional specification documents. They appeared in the body of the specification, directly supporting the definition of particular system components and functions. Ultimately, they appeared at the front of the document, conveying succinctly the design vision of the system: The traditional functional specification became the appendix. A survey of software designers in the mid-1980s, reported that scenarios were widely used as a working design representation (Rosson, Maass, Kellogg 88). (See also Wexelblat 87).

In the 1990s, the emergence of scenarios in the design of computers and software continued (see Carroll 95): Product planners presented day-in-the-life scenarios to their managers as design proposals (Dubberly, Mitsch 92; Erickson 95). Requirements engineers gathered workplace scenarios through direct observation and interviews, and analyzed scenarios as primary data (Antón, McCracken and Potts 94; Holbrook 90; Hsia et al. 94; Kaindl 97; Kuutti 95; Potts 95). Scenarios were used to specify functionality (Carroll and Rosson 90; Johnson and Benner 93). Scenarios were as employed as shared representations in design meetings, notably in meetings involving direct collaboration among designers and users (Karat, Bennett 91; Kyng 95; Muller et al. 95). Even the design of the underlying system software was addressed scenario by scenario (Jacobson, 1995; Jacobson, Booch and Rumbaugh 98; Kazman et al. 96; Rosson, Carroll 95; Wirfs-Brock 1995). Typical interaction scenarios were used as rubrics for system documentation, help, and user training (Carroll 90; Carroll 98; Karat 95), as well as for designing evaluation tasks (Carroll et al. 93; Nielsen 95). And scenarios were employed to codify and convey design knowledge among designers (Carey et al. 91; Young and Barnard 87).

Scenarios are now quite pervasive as a working representation in the design of computer systems and applications. Major design methods for human-computer interaction are scenario based (Beyer and Holtzblatt 98; Carroll 00; Constantine and Lockwood 99; McGraw and Harbison 97). Broader methodological frameworks for integrating human-computer interaction design with software engineering and strategic planning are based upon scenarios (Jarke, Bui and Carroll 98). The ISO 9241-11 international usability standard requires explicit analysis of context of use as part of the system design process (International Standards Organization 98), and scenarios are the chief candidate for this role (Dzida and Freitag 98). A recent survey of industrial software development groups reported that *all* groups surveyed employed scenario-based design methods — most had adopted scenario-based methods because of deficiencies of functional specification-based methods, and most desired design tools and techniques that more strongly scenario-based (Weidenhaupt et al. 98).

This was a remarkable evolution to experience. When I joined the IBM Corporation in 1976, no one spoke about design scenarios. We were caught up in the zeitgeist of the "new design methods" (Jones 70). This view of design promoted formal, or at least disciplined, representations and operations. It sought to move beyond the vagaries of craft design to a more scientific foundation. Document-driven design methods, incorporating functional specification documents, sprang from this zeitgeist. As I finish writing this chapter (in spring of 2000), I am on sabbatical at the Xerox Research Centre Europe in Cambridge, England: *Every* design discussion is about interaction scenarios.

2 A strange aid to thought

In this chapter, I consider why scenarios became so pre-eminent in software design. Interestingly, the main historical root emanates from strategic planning, not design studies; perhaps this is the legacy of the new design methods. Also interestingly, the foundations and rationale for scenario-based design, though they are substantively cognitive, do not originate in cognitive science work on problem solving. They largely pre-date cognitive science, and have developed independently of it. Nevertheless, it is interesting to reconstruct cognitive science foundations for scenario-based design. This could guide theory-based development of further tools and for techniques that might be overlooked on a purely practice-based understanding of support scenario-based design. It could also help to produce a more generalized understanding of scenario-based design that might facilitate its application in other design areas.

Tanaka (Tanaka 94) recalls that scenario-based approaches to strategic planning originated in work carried out in the Rand Corporation in the late 1940s. The best-known example of this work is in Herman Kahn's *Thinking about the unthinkable* (Kahn 62). For example, Kahn considered the Accidental War scenario (p. 145-150). In this scenario, an incident of equipment malfunction or unauthorized behavior results in the launch of a single Soviet missile. The missile detonates in Western Europe, and this is immediately detected and disseminated throughout western military and civilian installations. Although the incident is not interpretable as an attack, the level of anxiety throughout the Strategic Air Command results in one officer misunderstanding or disobeying orders, and firing the missiles under his command. The counter-attack is also not interpretable as an all-out first strike, however, it is an escalated military response to the original mistake, and could provoke the Soviets to launch their own ready missiles and bombers. In response, the U.S. might well order the rest of its missiles to be fired, as well as launching its bombers to protect them from the Soviet assault and to position them for a subsequent response.

Not a nice story. Indeed, Kahn's planning scenario became a shared nightmare for much of the world for a quarter-century. It was also a very useful design aid. It is still shocking to know that in early 1961 no one had considered planning to communicate with the U.S.S.R. during the five to forty minutes it would take for accidentally-launched missiles to reach their targets. And yet doing so could pretty literally save the world. This possibility was identified by working through the Accidental War scenario. Kahn also developed extensions of this scenario in which the U.S. and U.S.S.R. negotiated a limited nuclear war (this was later explored in the 60s movies "Fail-Safe" and "Dr. Strangelove") and in which they unilaterally established a world government (after surveying the first 10 million casualties of Armageddon).

Kahn called scenarios "a strange aid to thought," because they were not conventionally employed. However, he argued that scenarios provided five distinct advantages to strategic planners (p. 143-144). (1) Scenarios help analysts avoid the tempting assumption that circumstances will remain largely the same. They dramatically and persuasively emphasize the wide range of possibilities that must be considered. (2) Scenarios force analysts to address contextual details and temporal dynamics in problems, which they can avoid if they focus only on abstract descriptions. (3) By concretely illustrating a complex space of possibilities, and imposing a simple linear rubric of time, scenarios help analysts deal with several aspects of a problem situation simultaneously. They facilitate the comprehension and integration of many interacting elements — psychological, social, political, and military. (4) Scenarios stimulate the imagination, helping analysts to consider contingencies they might otherwise overlook. They vividly illustrate principles or issues that might be overlooked if one only considered actual events. (5) Scenarios can be used to consider alternative outcomes to past and present crises.

The objective of strategic planning is to anticipate and assess future contingencies, resultant situations, and the courses of action they would present. The planner analyzes the problematic nature of current and future situations, situations that are at best only partially understood, and describes and evaluates possible responses. This is actually very similar to design, in which the objective is to envision future artifacts, and to assess situations in which those artifacts would be employed in terms of personal and organizational consequences. Essentially, strategic planners design courses of action to meet the requirements of future contingencies.

3 The solution-first strategy

Planning and designing are both examples of ill-structured problem solving. In *Cognition and thought*, Reitman (Reitman 65) distinguished ill-structured problems from well-structured problems, such as the familiar puzzles investigated in psychological studies of problem solving, in three ways: In ill-structured problems, the problem is not fully defined; the possible moves are not all given; and the goal is unknown. The indeterminacy of ill-structured problems encourages a solution-first strategy. Thus, designers typically discover the nature of the problem they are solving by designing a solution and analyzing it. In this way, they exploit the concreteness of their own solution proposals to evoke further requirements for analysis. A key element in a solution-first approach is that designers typically need to abandon their own prior work when it has served the purpose of clarifying the problem. For example, we studied a librarian and systems engineer designing access to reference materials (Carroll, Thomas and Malhotra 79). They posed and solved half a dozen design problems in the course of clarifying the overall problem they were trying to solve.

At that point, they abandoned all the prior subsolutions, and quickly produced a comprehensive design.

Solution-first strategy entrains premature design action
Scenarios evoke reflection in design
Scenarios facilitate move testing in the context of design
Scenarios evoke empathy for actors, raise questions and concerns
Scenarios emphasize interactions and dynamics in design problems
Scenarios make problem framing feel like concrete work
Solution-first strategy entrains premature commitment to an initial solution
Scenarios relax commitment but support progress
Scenarios are concrete but sketchy, easily revised or replaced
Scenarios are vivid and incomplete, and therefore very evocative
Scenarios emphasize that circumstances are ephemeral
Solution-first strategy entrains insufficient analysis
Scenarios support many needs and purposes
Scenarios raise questions at many levels, of many types
Scenarios can be elaborated and analyzed in many ways
Scenarios help to integrate diverse partial solutions and views of a design
Solution first strategy entrains insufficient consideration of alternatives
Scenarios can be abstracted and categorized
Scenarios encourage grounded categorization and abstraction
Scenarios organize design knowledge in terms of tasks and artifacts
Solution-first strategy entrains inappropriate reuse of familiar approaches
Scenarios promote work-orientation
Scenarios discourage inappropriate reuse by focusing on tasks to be supported
Scenarios facilitate direct user involvement to ensure focus on real tasks
Scenarios emphasize that the user's experience is the ultimate object of design

Figure 1: Challenges of the solution-first strategy and properties of scenarios that may address them

The solution-first approach can be exciting, effective, and efficient, but it also entrains several well-known hazards. Designers (or planners) may generate partial solutions too quickly, that is, before sufficiently analyzing what is already known about the problem and the possible moves. This would tend to result in less valuable partial solutions, or even in partial solutions that obscure key pieces of relevant information. Designers may too readily reuse a solution approach they have employed before, one that is familiar and accessible, but perhaps not appropriate. They may generate an insufficient variety of partial solutions and thereby insufficiently explore the problem space. They may insufficiently analyze their own partial solutions; discovering some, but not all of the critical requirements. They may not be able to integrate partial solutions. Or they may have trouble abandoning an initial solution approach that has served its purpose (see Cross, this volume, for an extensive review).

Scenarios seem well-suited to support solution-first approaches in design, while at the same time mitigating its entrained hazards. In strategic planning and in the design of interactive systems and software, scenarios can be used to describe proposed solutions and their immediate consequences. Scenarios are easy to create, simple, and concrete; a designer can use them to quickly contrast alternative solution approaches. Scenarios are sketchy; their form alone suggests they are tentative and incomplete, but they are also vivid. This makes them evocative, but also helps differentiate them from "real" solutions. Since scenarios are merely envisionments, they can't be confused with implemented solutions, and are easily revised or abandoned. They provide task-oriented overviews of system functionality, for example, serving as "advance organizers" to functional specifications (Ausubel 60). Figure 1 presents a view of the hazards of solution-first design against the purported advantages of scenarios as aids to design thinking.

Working with scenario descriptions evokes reflection in the context of design activity, helping designers coordinate design action and reflection. They evoke empathy for the participants in a situation of use, raise questions for designers to address, and emphasize the dynamic flow of activity in the use of a designed artifact or system. Scenarios are concrete and flexible, they concretely fix an interpretation, but are open-ended and easily revised. This helps designers manage the fluidity of design situations — both by helping designers to avoid premature closure in their design work and by evoking new ideas to complete the scenario envisionment. Scenarios can be written at multiple levels, from many perspectives, and for many purposes. They afford multiple views of an interaction, diverse kinds and amounts of detailing, helping designers manage the many consequences entailed by any given design move. They encourage analysis and guide the integration of partial solutions.

Scenarios can also be abstracted and categorized, helping designers to recognize, capture, and reuse generalizations, and to address the challenge that technical knowledge often lags the needs of technical design. Domain-grounded abstractions help designers generate and consider alternatives. Finally, scenarios promote work-oriented communication among stakeholders. They anchor design discussion in work supporting participation among stakeholders and appropriate design outcomes, helping to make design activities more accessible to the great variety of expertise that can contribute to design, and addressing the challenge that external constraints on designers and clients often distract attention from the needs and concerns of the people who will use the technology.

4 Scenarios evoke reflection in design

Designers are intelligent people performing complex and open-ended tasks. They want to reflect on their activities, and they routinely do reflect on their activities. However, people take pride not only in what they know and learn, but in what they can do and in what they actually produce. There is a fundamental tension between thinking and doing: thinking impedes progress in doing, and doing obstructs thinking. Sometimes this conflict is quite sharp, as when one must stop and think before taking another step. But frequently it is more a matter of trading off priorities.

Donald Schön (Schön 83; Schön 87) has discussed this conflict extensively in his books on reflective practice. For example, he analyzes a coach reacting to an architecture student's design concept for a school building; the design includes a spiral ramp intended to maintain openness while breaking up lines of sight (she calls the idea "a Guggenheim"): "... when I visited open schools, the one thing they complained about was the warehouse quality — of being able to see for miles. It [the ramp] would visually and acoustically break up the volume" (Schön 87, p. 129).

The coach feels that she needs to explore and develop her concept more thoroughly, noting that a ramp has thickness and that this will limit her plans to use the space underneath the ramp; he urges her to draw sections. However, he does not bother to justify this advice to the student, to allow her to see the rationale behind his advice; as Schön puts it, he does not reveal "the meanings underlying his questions" (Schön 87, p. 132). Schön regards this as a hopeless confrontation in which no progress can be made on the particular design project, or on the larger project of understanding how to design. Both the student and the coach are willing to act publicly and to share actions, but do not reflect enough on their own and one another's values and objectives, and on their interpersonal dynamics.

Reflection is not always comfortable; it forces one to consider one's own competence, to open oneself to the possibility of being wrong. As Schön (87) put it, "The interactions I have suggested emphasize surfacing private attributions for public testing, giving directly observable data for one's judgments, revealing the private dilemmas with which one is grappling, actively exploring the other's meaning, and inviting the other's confrontation of one's own" (p. 141).

In general, people do not like to make themselves self-aware of their own roles as actors in situations; indeed, being "objectively self-aware" impairs performance of nonroutine tasks (Duval and Wicklund 72) — like designing software!

Designers do try to create opportunities for their own reflection. They organize design review meetings in which the whole team works through a set of requirements, a progress report, or a specification. It is also common to build early prototypes to verify and refine design requirements; one can directly observe prospective users interacting with such a prototype to make a formative evaluation of the design (Scriven 67). These activities can facilitate the identification and integration of different perspectives; they can raise concrete and detailed design issues to guide further work. In this way they help designers to reflect on the work they've already done. However, they do not evoke reflection *in the context of doing design*. Though design reviews and formative evaluations are reflective activities, they are ancillary activities that must be coordinated with design itself. Prototyping is directed at building and testing software that embodies a design, not on thinking about the design as it is produced.

Constructing scenarios of use helps to evoke reflection in the context of doing design. Consider the scenario in Figure Two: it succinctly and concretely conveys a problematic situation, in this case, a middle-school student is troubled by unresolved questions at the end of a class period. Her learning experience failed to fit into the time allotted. The teacher would like to help, but has few options; students are already arriving for the next science class. The teacher makes a generic suggestion, one that she probably knows will not be effective.

- Marissa was not satisfied with her class today on gravitation and planetary motion. She is not certain whether smaller planets always move faster, or how a larger and denser sun would alter the possibilities for solar systems.
- She stays after class to speak with Ms. Gould, but she isn't able to pose these questions clearly, so Ms. Gould suggests that she re-read the textbook and promises more discussion tomorrow.

Figure 2: Marissa's gravity project

The scenario raises many questions about the problem situation: Where are the other children? Why can't they just begin investigating Marissa's questions? Will those questions mean as much to Marissa tomorrow as they do at this moment? Why is it so critical that Marissa be able to explain everything to Ms. Gould? How will re-reading the text help? How could the situation be improved? Such a scenario can be regarded as a part of requirements analysis. In Kahn's sense, it forces the analyst to focus on the contextual details of the design problem.

Scenarios are effective in part because the human mind is especially adept at overloading meaning in narrative structures, both in generation and interpretation, as illustrated by the remarkable examples of dreams (Freud 00), myths (Levi-Strauss 67), and folktales (Propp 57). For example, when people communicate, they follow a convention that has been called the *given-new contract* (Haviland and Clark 74). They first summarize or allude to relevant background information, and then present what is novel. This normative structure eases both the generation and interpretation of narratives.

Schön (83, p. 67-68) drew an important contrast between merely creating or identifying elements of the problem context, and "giving them reason." Designers, as experts, will often have a category or a label with which to "understand" a situation. In reality, this simple step often establishes an insurmountable obstacle to achieving a real understanding. "When someone reflects-in-action, he becomes a researcher in a practice context." (p. 68)

A state of subjective self-awareness is encouraged by designing scenarios because the focus of attention is the activities and experiences of the prospective user.

Designers want to reflect, but they know from experience that it is impossible to fathom all the consequences and interdependencies. At the same time, they know that they can act and immediately see progress toward a solution, or at the least feel that they have eliminated some possibilities. Faced with this choice, it is always more attractive in the short-term to act. Creating and elaborating scenarios is concrete design work; the designer sees and feels progress toward a design result. At the same time, scenarios are concrete hypotheses about what the people using the design result will do, think and experience. Thus, in Schön's (83) terminology, scenarios evoke reflection-in-action. Schön stressed the importance for designers to experience the "felt-path" of the people interacting with their designs. A scenario guarantees this experience: it presents a potential felt-path to the designer; it is a medium through which designers can envision and explore alternative felt-paths.

5 Scenarios relax commitment but support progress

Design analysis is always indeterminate, because design changes the world within which people act and experience. The rapid evolution of spreadsheet software in the 1980s does not indicate a failure in the original requirements analysis for VisiCalc, but rather suggests the extent to which the original spreadsheet programs altered the work situations in which these program were used (Carroll and Rosson 91). Requirements always change (Brooks 95). When designs incorporate rapidly-evolving technologies, requirements change even more rapidly. The more successful, the more widely-adopted and the more impactful a design is, the less possible it will be to determine its correct design requirements. And in any case, refinements in software technology and new perceived opportunities and requirements propel a new generation of designs every 2-3 years.

There is a tendency to think of the indeterminacies of technology development in positive terms: the world is getting better, albeit sometimes in ways we didn't expect. Today this positive attitude is balanced with growing acknowledgment of potential negative consequences, such as pollution of groundwater and the deterioration of the atmosphere. But both these emphases underanalyze the extent to which design, and especially the design of new technology, *undermines the stability of the world.* Schön (67), for example, noted that in our era technology development constantly erodes the manager's concept of "the business I'm in" (p. 195) and a worker's notion of the special skill he or she possesses; it erodes the traditional staging of life into education followed by steady practice. Even writers, such as McLuhan (94) who revel in the prospect of "learning a living," agree about the magnitude of this instability.

The instability of design situations originates inside design teams as well as outside. As a project goes forward, its funding may be threatened or restructured; various stakeholders or team members may change their interests or priorities, or may even leave the team. Others with unknown interests and priorities may join. This creates a chronic need for education and consensus-building to ensure that there is agreement as to what the requirements are at any given point in time.

Schön (67) stressed that in the face of great instability, people create illusions of stability to manage their own uncertainties and potentially disturbing perceptions. This is consistent with a substantial body of social psychology research: People create logically tidy interpretations of their experience, even if they must "adjust" their actual perceptions in order to do this (Nisbett and Wilson 97). People protect their interpretations by selectively reconstructing their own experience (Erikson 1980). And the more reconstructionist effort people must expend to maintain an interpretation, the more ardently it is held, even in the face of incontestable disconfirmation (Festinger 57).

Given the human bias for stability, we should not be surprised by the persistence of the attitude that design problems can be planfully decomposed into routine subproblems, that scientific principles can be mechanically applied to expose and manage the underlying orderliness of design problems. From the standpoint of the psychology of designers and managers, it could conceivably become even stronger as it becomes more clearly inappropriate (Festinger, Riecken and Schachter 56). These dysfunctional beliefs about design would not be such a problem if they could only lead to standstill. The potentially dangerous aspect is that they can lead to "success" in accomplishing the wrong things. Designers almost always design something. If their need for stability induces them to prematurely close their designs, to cease reflecting and critiquing, to declare victory, they may produce a solution for requirements that no longer exist.

To manage an ambiguous and dynamic situation, one must be concrete but flexible. One must be concrete just to avoid being swallowed by the indeterminacies; one must be flexible to avoid being captured by a false step. Systematic decomposition is a traditional approach to managing ambiguity, but it does not afford flexibility. Instead one ends up with a set of concrete subsolutions, each of which is fully specified. Unfortunately, by the time the set of subsolutions is specified, the requirements often have changed.

Scenarios of use reconcile concreteness and flexibility. They are concrete in the sense that they simultaneously fix an interpretation of the design situation and offer a specific solution: the scenario in Figure 3 specifies a usage experience that addresses some of the problems raised in Figure 2, and that could be prototyped and tested. At the same time the scenario is flexible, deliberately incomplete and easily revised or elaborated: in a few minutes, a piece of the scenario could be re-written (e.g. reconsidering what notebook page types are necessary and desirable) or elaborated (e.g. further specifying process of collaborative writing or the annotation functionality).

- Marissa was not satisfied with her class today on gravitation and planetary motion. She is not certain whether smaller planets always move faster, or how a larger and denser sun would alter the possibilities for solar systems.
- Through the network-viewer client on her personal computer, she accesses the county virtual shared-school server. She is presented with a control panel incorporating a list of her current project groups and ad hoc groups open to all, and indicating the activity-status of each.
- She opens the shared notebook, chat, and video conferencing tools associated with one of the "open" groups. Randy and David (Randy is at his home, David is in the high school physics lab) are already working together on whiteboard page in the notebook with the Alternate Reality Kit, a science object that allows students to alter various physical parameters (like the universal gravitational constant) and observe the effects in a simulated world.
- The three students discuss possible experiments, and then build and analyze several solar systems, eventually focusing on the question of how massive comets might disrupt otherwise stable planetary systems.
- They capture and visualize data from these experiments in a data analysis page, and together write a brief report as an introduction to their notebook. They send email to Don, a student in Randy's school, and Ms. Gould, asking them to review and annotate the project notebook.

Figure 3: Marissa's gravity project

Concrete material is cognitively accessed and interpreted more easily and more thoroughly than abstract material. For example, people can remember a prototypical instance far better than they can remember the definition of the category to which that instance belongs (Medin and Schaffer 78; Rosch et al. 76). Incomplete material tends to be elaborated with respect to one's own knowledge when it is encountered. This process of elaboration creates more robust and accessible memories, relative to memories for more complete material (Wertheimer 38). The combination of concreteness and incompleteness in scenarios engages a powerful variety of constructive cognitive processes.

A scenario provides a concrete envisionment of a design solution, but can be couched at many levels of detail. Initial scenarios are typically very rough. They specify a system design by specifying the tasks users can or must carry out, but without committing to the lower-level details of precisely *how* the tasks will be carried out or *how* the system will enable the functionality for those tasks. In Kahn's sense, this can help remind the designer that many possibilities and contingencies should and can be considered. The narrative in Figure 3 is at an intermediate level, with some detail regarding task flow, but not at the level of individual user-system interactions. In a scenario-based design process, designers develop a variety of alternate detailings of descriptions like Figures 2 and 3.

Thus scenarios provide a rich foundation for action-oriented reflection. By being both concrete and rough, they make explicit the design goal of specifying tasks and functions in greater detail. But they do this by opening up the issue of how to do this detail design, instead of closing it out prematurely. Scenarios allow designers to provisionally construct a space of user tasks despite the instability in requirements originating from the context of technology development. Scenarios are particularly appropriate for managing the instability originating inside design teams: They are broadly accessible to various stakeholders and team members, unlike artifact-oriented decompositions such as functional specifications.

The notion that scenarios of use are an appropriately concrete but rough design object extends several earlier analyses. For example, Schön (83, p. 277-279) emphasized that effective reflection must be tightly coupled to action: the analysis need not be complete and consistent, it need only guide a restructuring of the current situation that can produce new design actions or new insights. His cases are drawn from design domains for which there are rich languages to allow designers to quickly create and explore situations. Scenarios are such a language for the design of human-computer interactions.

In an earlier book, Schön (67, p. 41) argued that decomposition is useful chiefly through the side-effects of eliciting and focusing concrete action, and thereby reducing uncertainty. But he warned that the resultant decomposition can "strangle" innovation. Ackoff (79) similarly developed the notion of "idealized design": a process of planning "the system with which the designers would replace the existing system *now* if they were free to do so" (p. 191). He argued that such a process can increase participation among stakeholders, mobilize participants with respect to the project, expand the concept of what is feasible, increase both creativity and the scope of human and organizational consequences that are considered in the design process; an idealized design is really just a vehicle for focusing the articulation of possibilities and the discussion of their consequences.

The fluidity of design situations demands that solutions be provisional, that commitments be tentative; yet if every design decision is suspended, the result will be a design space, not a design. A scenario is a concrete design proposal that a designer can evaluate and develop, but is also rough in that it can be easily altered and allows many details to be deferred.

6 Scenarios support many needs and purposes

Every element of a design, every move that a designer makes, has a variety of potential consequences. In their work on violin design, the Catgut Society found that additional rib holes were needed in their new treble violin to pitch the instrument high enough and yet maintain the neck length for fingering. This design move had the intended consequences, but also forced the designers to consider stronger materials for the ribs, and ultimately to use *aluminum* instead of wood. However the aluminum ribs caused a nasal quality in the lower strings, forcing the designers to reconsider a new wooden rib design — ribs so thin (0.7 mm) that the designers believed on analytical grounds that they would collapse. A typical complication in this design case is that the various consequences belonged to different stakeholders: A fingerboard that is too short is a problem for the musician. A very thin wooden rib is a problem for the instrument maker. A nasal tone is a problem for the musician and listener (Hutchins 81).

Schön (83, p. 101) sees design as a "conversation" with a situation comprised of many interdependent elements. The designer makes moves and then "listens" to the design situation to understand their consequences:

> "In the designer's conversation with the materials of his design, he can never make a move which has only the effects intended for it. His materials are continually talking back to him, causing him to apprehend unanticipated problems and potentials."
>
> Thus, the Catgut group made the move of employing aluminum ribs and then noticed the unanticipated problem of a nasal tone in the lower strings. When a move produces unexpected consequences, and particularly when it produces undesirable consequences, the designer articulates "the theory implicit in the move, criticizes it, restructures it, and tests the new theory by inventing a move consistent with it" (Schön 83, p. 155).

Schön's approach to managing the interdependencies within a design situation by treating design as inquiry raises several questions. What directs the designer to investigate various types of consequences, to listen for various types of backtalk? How are different types and sources of backtalk integrated in the designer's restructured design theory? The moves for the treble violin had various consequences. The Catgut group effectively integrated these in their subsequent design work, but how did they do that and, more generally, how can we support outcomes like that? Designers need a language for this conversation and they need techniques for managing the consequences.

Scenarios of use are multifarious design objects; they can describe designs at multiple levels of detail and with respect to multiple perspectives. They can help designers deal with several aspects of a problem situation simultaneously. The scenarios in Figures 2 and three provide a high-level task view, but easily could be elaborated with respect to the detailed moment-to-moment thoughts and experiences of the people in order to provide a more detailed cognitive view, or in terms of moment-to-moment actions to provide a more detailed functional view. Figure 3 could be elaborated in terms of the hardware and software components that could implement the envisioned functionality in order to provide a system view. Use cases play this role in object-oriented software engineering (Jacobson 95; Wirfs-Brock 95). Each of these variations in resolution and perspective is a permutation of a single underlying scenario. Indeed, the permutations are integrated through their roles as complementary views of the same design object.

Scenarios can leave implicit the underlying causal relationships among the entities in a situation of use. In Figure 3, the envisioned speech annotation capability allows the user to add a personal comment without the overhead of opening an editor and typing text. However, the annotation is noncoded, and thus cannot be edited symbolically. These tradeoffs are important to the scenario, but often it is enough to imply them (this is an aspect of the roughness property discussed above).

There are times, however, when it is useful to make these relationships explicit. For example, in another situation Marissa and her friends might want to study the disintegration of a moon in an unstable orbit. It could turn out that their version of the Alternate Reality Kit does not model such phenomena, or that it produces a simulation result that it physically inaccurate for the parameters they wish to investigate. Thus, this new scenario would end in failure. To understand, address, and track the variety of desirable and undesirable consequences of the original annotation design move, the designer might want to make explicit the relevant causal relationships in the scenario. Doing so provides yet another view of the envisioned situation as shown in Figure 4.

Concrete virtual objects, interactive simulations
- + afford manipulation and encourage analysis
- + provide access to phenomena not available in the real world
- - have limitations, boundary conditions, may break down
- - concepts and skills that are learned may not generalize

Direct connections from simulation to analysis tools
- + are convenient, allow rapid and fluid investigation
- - limit learner control and insight into what is happening

Cumulative recording of experiments
- + encourages reflection and analysis
- - creates additional bookkeeping work
- - limits flexibility, makes learning too product-oriented

Figure 4: Marissa's gravity project as consequences associated with envisioned features of the virtual school.

Scenarios can help designers move more deliberately toward specific consequences. For example, it might be decided to provide explicit visual programming control of the connection between simulation and analysis tools, but to also allow user-defined macros to support rapid and fluid investigation for those students who have mastered basic relationships. Alternatively, analysis tools might be designed as agents that take initiative in offering to process data appropriate for them. Each of these alternatives would entrain different elaborations for both various virtual school scenarios. These elaborations could then be explored for further consequences and interdependencies.

Scenarios can be interpreted at different levels. For example, the scenario in Figure 3 can be viewed in aggregate, raising organizational questions for the county school system, instead of individual learning-activity questions. For example, is there enough network bandwidth to support this vision? Is the pedagogy effective; will Marissa and her friends still do well on the state's standardized achievement tests? What is the teacher's role in this educational paradigm; will teachers have to be retrained?

Using scenarios in this way makes them a more powerful design representation. They allow the designer the flexibility to develop some use scenarios in great detail, for example the ones that describe the core application functionality or the critical usability challenges, while merely sketching other less problematic scenarios. At the same time, they allow the designer to switch among scenario perspectives and to directly integrate, for example, usability views with system and software views. Such a flexible and integrative design object can help designers manage the nexus of consequences entrained by design moves.

The interconnectedness of design decisions, and the variety and extent of any given decision's consequences, requires designers to consider their decisions from many different perspectives: software architecture, marketing, ease of learning, production cost, usability, and so forth. It requires them to consider their decisions at many different levels of detail in the proposed solutions. Scenarios of use serve as a concrete context for developing and integrating these different perspectives and levels.

7 Scenarios can be abstracted and categorized

Schön (83, p. 42-44) emphasized the "dilemma" of rigor or relevance throughout the professions. He describes the "high, hard ground" where practitioners can make use of systematic methods and scientific theories, but can only address problems of relatively limited importance to clients or to society. He contrasts this to the "swampy lowland [of] situations ... incapable of technical solution" (p. 42); here the practitioner can confront the greatest human concerns, but only by compromising on technical rigor. The design and development of technology aspires to occupy the high, hard ground, to apply the current state of technical knowledge systematically, to reduce difficult problems to mere corollaries — and of course to bask in the confidence that deductive science confers both on its practitioners and the recipients of their efforts: science is certainty. But at the same time, technology design and development is inevitably driven to pursue novelty and innovation, and thereby to slip continuingly into the swampy unknown regions.

Schön (83, p. 16, 42-44) discusses the field of operations research as an example. During World War II, operations research developed from the successful application of mathematics to modeling battle situations like submarine search. After the war, the field expanded its interests to management problems in general, and successfully produced effective formal models for relatively bounded problems like managing capital equipment replacement or investment portfolios. However, it generally failed in complex, less well-defined areas like business management, housing policy, or criminal justice. Remarkably, the response of the field to this was largely to focus attention on the relatively simple problems that suited the mathematical models.

Though technical design cannot escape Schön's swamp, designers need to extract lessons from their experience to guide their work and improve their practice. If technical knowledge lags design, then designers themselves must formulate what they learn — perhaps becoming creators of technical knowledge more than consumers of it. The Catgut Society is a beacon for this: At the start of their project there was little relevant technical knowledge; physicists had a good understanding of vibrating strings, but no significant understanding of vibrating systems composed of strings vibrating across air cavities partitioned by wooden ribs and bounded by irregularly-shaped wooden boxes. The Catgut work in the end has substantially advanced technical knowledge of complex acoustical systems: the design work created an "island" of hard ground in the swamp of real problems.

Scenarios keep the designer of computer systems and applications in the swamp, but by their very nature also provide scaffolding to get a view of the design situation from a bit higher up. The roughness of scenarios is, after all, abstractness: To the extent that a scenario is rough, it is a description of a space of particular user interactions. This is analogous to the way in which a jazz score is a rough description of the many ways in which that piece of music might be performed. Documenting and explaining a scenario provides an account of a class of possible situations; this creates a generalized design object that can be immediately employed in further design work by being elaborated in various ways. For example, the scenario in Figure 3 could guide the design of a multitude of information systems.

But this is only the most primitive technique for developing design knowledge with scenarios. Scenarios exemplify particular themes and concerns in work and activity situations. Earlier we discussed two educational scenarios. In one (Figure 2), the learner is unable to pursue a current concern, in the other (Figure 3), a learner has several options available and chooses to pursue one (perhaps because it matches a personal concern). Described at this level of generality, these are not scenarios unique to educational systems. They are general patterns for how people work pursue goals. Thus, it is possible that some of the lessons learned in these situations could be generalized to other situations in the same categories. A taxonomy of scenarios provides a framework for developing technical design knowledge.

Scenarios can also be classified in terms of the causal relations they comprise. In Figure 3, for example, providing direct piping from the simulation to the data analysis tools simplifies the interaction. In this causal relation, the consequence is the simplification of syntactically specifying the input to the analysis tools. Generalizing the relation in this way allows the feature associated with the consequence (in this example, direct piping) to be understood as a potential means for that consequence, and employed to that end in other design contexts. There is of course no guarantee that the generalization is correct, that can only be settled by trying to use it and succeeding or failing. The point is that such candidate generalizations can be developed from scenario descriptions.

Scenarios can also be taken as exemplars of model scenarios for example, Figure 3 illustrates a model of direct engagement (Hutchins, Hollan and Norman 86). The students interact with the objects of a simulation directly, moving and otherwise manipulating them. Other models are possible of course; the interaction might have been command-driven, or might have required the students to learn a modeling representation. Seeing the scenario in Figure 3 as a direct engagement scenario allows the designer to benefit from prior knowledge pertaining to this model and to contribute further design knowledge of the model based on the current project.

Designers are not just making things; they are making sense. Particularly in technical design it is not enough for them to master a craft practice, for there is no stable craft practice in design domains like computer systems and applications. The turbulence of technology development leaves little unchanged, but particularly in the short run, leaves little resolved. We may expect that conventional science will eventually systematize technical knowledge in new domains, but we also know that it typically will do so after the wave of technological innovation has swept onwards. The challenge for designers is to learn as they go.

Technical innovation is driven to those regions where technical knowledge is thin. Designers often have little more than craft practices to guide them in these regions. But if we see design as inherently a process of inquiry, this lag between codified knowledge and practice is transformed into a significant opportunity. Design can be a paradigm for creating and cumulating technical knowledge. Scenarios provide a broad rubric for organizing and generalizing knowledge attained in design contexts. They can be abstracted and categorized in various ways, and then employed in new design problems.

8 Scenarios promote work-orientation

Designers must have constraints; there are just too many things that might be designed. Requirements, if they can be identified, are clearly the best source of constraints because they indicate what sort of design work is needed. But there are many other sources of constraints. The current state of technology development makes some solutions impossible and others irresistible: On the one hand, designers cannot use technology that does not yet exist, though their work often drives technology development toward possibilities that are nearly within reach. On the other hand, designers, like everyone else, are caught up in a technological zeitgeist that biases them toward making use of the latest gadgets and gizmos. In addition, designers are often biased toward deploying technologies they have used before, even when they are aware of limitations in these technologies.

As Churchman (70) observes, many systems are designed to serve the "wrong" client, for example, hospitals are designed to serve doctors, not patients. In time, of course, a system designed to serve the wrong client will poorly serve all clients. This is similar to the confusion of "customers" and "users" in identifying design requirements. However, by far the worst such confusion is entrained by solving tidy problems instead of addressing the real problem. In the systems approach to instructional design, the mistaken client is the instructional designer, or perhaps the instructional design firm; the underlying intent is to streamline the production of instruction, not to improve its utility.

Many constraints in design originate in the organizational structures within which design work is embedded. Only certain types of assumptions and arguments are acceptable in the business cases that underwrite design projects. For example, well into the 1980s it was widely believed that executives would never use a keyboard. This belief was based on a marketing stereotype of executive disdain for clerical tasks, yet it constrained many technical strategies for the development of computer hardware and software in that period. Ultimately, appropriate hardware and software were designed for executives, as one can easily verify in the number of notebook computers in use in the first class cabin of an airplane. Apparently, the issue had nothing to do with typing per se, but with other aspects of use situations.

Power structures and stereotypes play assorted other roles. Design projects are often chartered with a priori commitments to follow a strict decompositional approach: which makes them easy to manage, but unlikely to succeed with respect to serving the needs of users and clients. In such a context, designers must struggle not only with technical challenges but with an impossible methodology. Schedules and resources are often assigned in ways that create on-going conflicts between system designers and usability engineers: the usability engineers need to evaluate scenarios and prototypes at every stage of system development, but if schedules and resources do not provide for this, the usability work can conflict with development work.

In all these examples of external constraints, the designers can become distracted by ancillary or even perverse factors and lose sight of what is essential in the design project, namely, the needs and concerns of users. The designer can become "unsituated" with respect to the real design situation, which is not the marketing manager's projections, or the instructional designer's list of steps, or the software engineer's system decomposition. The real design situation is the situation that will be experienced by the user, and designers need to stay focused on that.

Scenarios are work-oriented design objects. They describe systems in terms of the work that users will try to do when they use those systems. A design process in which scenarios are employed as a central representation will ipso facto remain focused on the needs and concerns of users (Carroll and Rosson 90). Scenarios can address the so-called representational bias in human cognition, which inclines people to overestimate the relevance of things that are familiar to them (Kahneman and Tversky 72; Tversky and Kahneman 74). This tendency is extremely difficult to mitigate, but can be managed by making exceptional patterns vivid. Scenario narratives describing critical and exceptional circumstances can be a tool for managing this problem.

One can increase the effectiveness of scenarios of use as work-oriented design objects by couching them at an appropriate level and by directly involving users in creating them. A person's experience of working with a system revolves around understanding and developing task goals, responding opportunistically to intriguing system events, and planning, carrying out and evaluating courses of action. People do not focus on the very low-level enabling actions and events that comprise these experiences, such as keypresses, beeps, mouse gestures and clicks, display updates, and font size. Quite often people are only aware that they are seeing, hearing and doing these things when something goes amiss. A heuristic for writing scenarios focused on the client's needs and concerns is to initially couch them at the "basic task" level — the level at which people experience their own activity (Carroll and Rosson 91).

Ackoff (79) argued that the indeterminacy of design situations made it imperative that *all* stakeholders participate directly. Assuming that stakeholders can represent their own interests, this proposal pretty much guarantees an adequate focus on client's needs and concerns. However, it may not be the case that all stakeholders can represent their interests: they will be trying to evaluate their needs as transformed by new technologies that they may not understand. Moreover, it is not always feasible to include all stakeholders; in the design of a new spreadsheet application for personal computers there might be several million stakeholders.

Ackoff's ideas have been refined through various developments in participatory design over the past two decades. One technique is to include representative clients on the design team, instead of imagining that every client can participate directly. Of course client representatives on participatory design teams are not representative clients; by definition only clients not on the design team can be truly representative. However, given this dilemma, "representative" sampling may be a reasonable heuristic. Clients can also be helped to better represent their interests in design collaborations by taking part in facilitated demonstrations of possible scenarios of use (Kyng 95; Muller et al. 95).

Like every human activity, design work occurs in many overlapping social contexts: technical societies, corporations, states of technology development, industries, and so forth. Each context introduces constraints on possible methods and solutions. Sometimes these are constructive and appropriate, often they are neither. The cacophony of these social contexts and constraints can disorient designers with respect to their primary source of constraint — the needs and concerns of the people who will use the system being designed. Scenarios help keep the whole enterprise focused on past and future situations of use, they help keep designers focused on what matters most.

9 Some remarks on generalizing and teaching scenario-based design

It may be that scenario-based design has attained pre-eminence in the domain of computer systems and software because software is so malleable as a design material, and yet also so complex. It is particularly easy to create a result in software, and particularly difficult to analyze why it is wrong and how to fix it; it is particularly easy and particularly costly to engage in solution-first design. As interactive systems (most notably the PC), end-user applications (like spreadsheets and word processors), and graphical user interfaces (like that of the Apple Macintosh) became pervasive in the early 1980s, these characteristics became a driving requirement for design methods. I believe this is fundamentally why scenarios marched forward in functional requirements documents during the 1980s, finally becoming themselves the main content.

However, there is also something deeper going on. Scenario-based design ameliorates the problems entrained by solution-first design in such a variety of ways, it can seem like a happy coincidence of a bag of tricks. But in fact scenario-based design changes the object of design from an artifact or system in the world to a series of actions and experiences in human activity. This is perhaps why the deepest root of scenario-based design is in strategic planning instead of design studies. Scenario-based design is the design of plans for people and their organizations, the novel artifacts and systems entailed by these plans are significant only to the extent that they do effectively facilitate human activity.

How general an idea is scenario-based design? Would it be applicable and effective in the design of mechanical systems or buildings? Perhaps it is not an accident that the technique developed in strategic planning and software design. Most essentially, computer systems and software are what people can do with them, and with what consequences. Albeit after more than two decades of evolution, this mantra is generally accepted in the computing industry. But the underlying principle may not be as true in other design domains. This indeed could be evidence of a significant category boundary among types of ill-structured problems.

Alternatively, it could present an opportunity to broaden the application of scenarios in other design domains. For example, engineers typically analyze consequences of failure in test cases, but it is not standard to include analysis of the human systems in which physical systems are embedded. Architects often make reference to people moving through the spaces they create, but it is not typical to explicitly analyze tradeoffs in the social experience of spaces. In Schön's Guggenheim example, the student and coach articulate and interrogate design elements that have interrelated implications for people moving in the space, but neither integrates the design elements into a coherent projected experience of the space. They do not construct and analyze a scenario. The question of how scenario-based design may generalize to other design domains can only be satisfactorily addressed through the development of a taxonomy of design domains. But it can be clarified considerably just by trying out scenario-based techniques in other design domains.

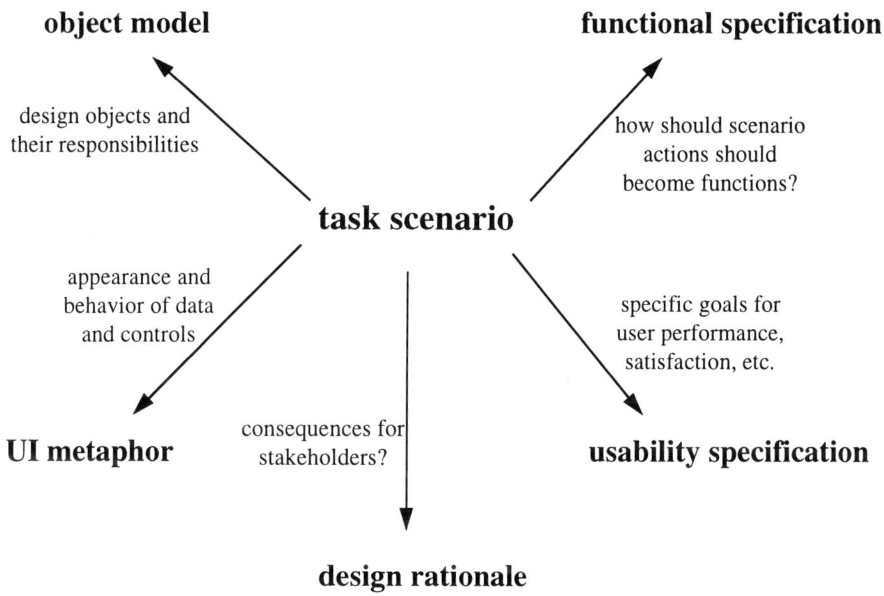

Figure 5: Views of task scenario within the design process

In recent courses for NASA, I used Figure 5 to convey the idea that scenarios integrate what could seem to be disparate design concerns, and that in doing so they enhance the possibility of reuse (albeit at a high level) within a project. The students in this course were mostly aerospace, mechanical and electrical engineers, though there were also some software engineers. I used the figure to emphasize that designing a concrete representation of future user interaction could be leveraged in a variety of system design and development activities.

This single visualization seemed to have a lot of impact on the students (as indexed by external signs of paying more attention and making notes). This has motivated me to create an even more ambitious version of the figure in which the scenario points also to formative evaluation, summative evaluation, system vision, and documentation, training and help (the last three grouped as one node; see Carroll 00). In my future teaching I want to investigate this particular lever for trying to arouse and focus the interest of professionals, and especially to continue to investigate whether non-software engineers perceive any utility for scenario-based design.

One NASA engineer in this class declared to me that the successor to the Space Shuttle will be a case study of scenario-based design. I hope it is. This would surely be the kind of broader design application we would need to begin to assess the generality of scenario-based design. For now, scenario-based design techniques will probably continue to develop chiefly within the computer industry.

10 Acknowledgements

This paper is a revised and extended version of "Five reasons for scenario-based design," which appeared in R.H. Sprague (Ed.), *Proceedings of the 32nd Annual Hawaii International Conference on System Sciences* (Maui, January 5-8, 1999). Los Alamitos, CA: IEEE Computer Society (published as CD-ROM). Thanks to Mike McCracken and Wendy Newstetter for helping to develop the version for this volume.

References

Ackoff R L, (1979), Resurrecting the future of operations research, *Journal of the Operations Research Society,* 30(3), 189-199.

Antón A, McCracken W. M, and Potts C, (1994), Goal decomposition and scenario analysis in business process reengineering, *Proceedings of CAiSE'94: Sixth conference on avanced iformation systems engineering,* (Utrecht, the Netherlands, June 6-10), Springer-Verlag.

Ausubel D R, (1960), The use of advance organizers in the learning and retention of meaningful verbal material, *Journal of Educational Psychology,* 51, 267-272.

Beyer H, and Holtzblatt K, (1998), *Contextual design: Defining customer-centered systems,* Morgan Kaufmann, San Francisco, CA.

Brooks F, (1995), *The Mythical man-month: Essays on software engineering,* Addison-Wesley, Reading, MA (Anniversary Edition, originally 1975).

Carroll J M, (2000), *Making use: Scenario-based design of human-computer interactions,* MIT Press, Cambridge, MA.

Carroll J M, (1990), *The Nurnberg funnel: Designing minimalist instruction for practical computer skill,* MIT Press, Cambridge, MA.

Carroll J M, (1994), Making *use* a design representation, *Communications of the ACM, 37*(12) 29-35.

Carroll J M, Ed. (1995), *Scenario-based design: Envisioning work and technology in system development,* John Wiley & Sons, New York, NY.

Carroll J M, Ed. (1998), *Minimalism beyond "The Nurnberg funnel,"* MIT Press, Cambridge, MA.

Carroll J M, Koenemann-Belliveau J, Rosson MB, and Singley MK, (1993), Critical incidents and critical threads in empirical usability evaluation, J Alty, D Diaper and S Guest, Eds. *People and computers VIII: Proceedings of HCI'93* (Loughborough, U.K., September), Cambridge University Press, Cambridge, MA, 279-292.

Carroll J M, and Rosson M B, (1985), Usability specifications as a tool in iterative development, HR Hartson, Ed. *Advances in human-computer interaction,* Ablex, Norwood, New Jersey.

Carroll J M, and Rosson M B, (1990), Human-computer interaction scenarios as a design representation, *Proceedings of the 23rd annual Hawaii international conference on systems sciences* (Kailua-Kona, HI, January 2-5), IEEE Computer society Press, Los Alamitos, CA, 555-561.

Carroll J M, and Rosson M B, (1991), Deliberated evolution: Stalking the view matcher in design space, *Human-Computer Interaction,* 6, 281-318.

Carroll J M, and Rosson M B, (1992), Getting around the task-artifact cycle: How to make claims and design by scenario, *ACM Transactions on Information Systems,* 10, 181-212.

Carroll J M, Thomas J C, and Malhotra A, (1979), A clinical-experimental analysis of design problem solving, *Design Studies,* 1, 84-92.

Churchman W, (1970). Operations research as a profession, *Management Science,* 17(2), 37-53.

Constantine L L, and Lockwood L A D, (1999), *Software for use: A practical guide to the models and methods of usage-centered design.* Addison-Wesley, Reading, MA.

Dubberly H, and Mitsch D, (1992), Knowledge navigator, BA Myers, Ed. *CHI'92 Special video program: Conference on human factors in computing systems,* (Monterey, CA, May 3-7), ACM SIGCHI, New York, NY.

Duval S, and Wicklund R A, (1972), *A theory of objective self-awareness*, Academic Press, New York, NY.

Dzida W, and Freitag R, (1998), Making use of scenarios for validating analysis and design, *IEEE Transactions on Software Engineering,* 24(12), 1182-1196.

Erickson T, (1995), Notes on design practice: Stories and prototypes as catalysts for communication, JM Carroll, Ed. *Scenario-based design: Envisioning work and technology in system development,* John Wiley & Sons, New York, NY, 37-58.

Erikson E H, (1980), *Identity and the life cycle*, Norton, New York, NY.

Festinger L, (1957), *A theory of cognitive dissonance,* Harper & Row, New York, NY.

Festinger L, Riecken H W, and Schachter S, (1956), *When prophecy fails*, University of Minnesota Press, Minneapolis, MN.

Freud S, (1900), *The interpretation of dreams, Standard Edition,* 4, Hogarth, London.

Go K, and Carroll J M, (Submitted), *Blind men and an elephant: Views of scenario-based system design.*

Haviland S E, and Clark H H, (1974), What's new? Acquiring new information as a process in comprehension, *Journal of Verbal Learning and Verbal Behavior,* 13, 512-521.

Holbrook H, (1990), A scenario-based methodology for conducting requirements elicitation, *ACM SIGSOFT Software Engineering Notes, 15(1),* January, 95-103.

Hsia P, Samuel J, Gao J, Kung D, Toyoshima Y, and Chen C, (1994), Formal approach to scenario analysis, *IEEE Software,* March, 33-41.

Hutchins C M, (1981), The acoustics of violin plates, *Scientific American,* 245(4), 170-186.

Hutchins E, Hollan J, and Norman D, (1986), Direct manipulation interfaces, D A Norman and SW Draper, Eds. *User centered system design,* Lawrence Erlbaum and Associates, Hillsdale, NJ, 87-124.

International Standards Organization, (1998), *ISO 9241-11, Ergonomic Requirements for Office Work with Visual Display Terminals (VDTs) — Guidance on Usability,* ISO, Geneva, Switzerland.

Jacobson I, (1995), The use-case construct in object-oriented software engineering, JM Carroll, Ed. *Scenario-based design: Envisioning work and technology in system development,* John Wiley & Sons, New York, NY, 309-336.

Jacobson I, Booch G, and Rumbaugh J, (1998), *The unified software development process*, Addison-Wesley, Reading, MA.

Jarke M, Bui X T, and Carroll J M, (1998), Scenario management: An interdisciplinary approach, *Requirements Engineering*, 3(3-4), 155-173.

Johnson W L, and Benner K M, (1993), Developing formal specifications from informal requirements, *IEEE Expert*, August, 82-90.

Jones J C, (1970), *Design methods: Seeds of human futures,* John Wiley & Sons, New York, NY.

Kahn H, (1962), *Thinking about the unthinkable*, Horizon Press, New York, NY.

Kahneman D, and Tversky A, (1972), Subjective probability: A judgement of representativeness, *Cognitive Psychology*, 3, 430-454.

Kaindl H, (1997), A practical approach to combining requirements definition and object-oriented analysis, *Annals of Software Engineering*, 3, 319-343.

Karat J, (1995), Scenario use in the design of a speech recognition system, JM Carroll, Ed. *Scenario-based design: Envisioning work and technology in system development,* John Wiley & Sons, New York, NY, 109-133.

Karat J, and Bennett JB, (1991), Using scenarios in design meetings – A case study example, J Karat, Ed. *Taking design seriously: Practical techniques for human-computer interaction design,* Academic Press, Boston, 63-94.

Kazman R, Abowd G, Bass L, and Clements P, (1996), Scenario-based analysis of software architecture, *IEEE Software*, November, 47-55.

Kuutti K, (1995), Work processes: Scenarios as a preliminary vocabulary, JM Carroll, Ed. *Scenario-based design: Envisioning work and technology in system development,* John Wiley & Sons, New York, NY, 19-36.

Kyng M, (1995), Creating contexts for design, JM Carroll, Ed. *Scenario-based design: Envisioning work and technology in system development,* John Wiley & Sons, New York, NY, 85-107.

Lévi-Strauss C, (1967), *Structural anthropology,* Anchor Books, Garden City, NY.

Medin D L, and Schaffer M M, (1978), A context theory of classification learning, *Psychological Review,* 85, 207-238.

McGraw K, and Harbison K, (1997), *User-centered requirements: The scenario-based engineering process,* Lawrence Erlbaum and Associates, Mahwah, NJ.

McLuhan M, (1994), *Understanding media: The extensions of man,* MIT Press, Cambridge, MA (original edition, 1964).

Muller M J, Tudor L G, Wildman D M, White E A, Root R A, Dayton T, Carr R, Diekmann B, and Dykstra-Erickson, E, (1995), Bifocal tools for scenarios and representations in participatory activities with users, JM Carroll, Ed. *Scenario-based design: Envisioning work and technology in system development,* John Wiley & Sons, New York, NY, 135-163.

Nielsen J, (1995), Scenarios in discount usability engineering, JM Carroll, Ed. *Scenario-based design: Envisioning work and technology in system development,* John Wiley & Sons, New York, NY, 59-83.

Nisbett R E, and Wilson T D, (1997), Telling more than we can know: Verbal reports on mental processes, *Psychological Review,* 84, 231-259.

Potts C, (1995), Using schematic scenarios to understand user needs, *DIS'95: ACM symposium on designing interactive systems*, ACM Press, New York, NY, 247-256

Propp V, (1958), *Morphology of the folktale*, Mouton, The Hague (originally published in 1928).

Reitman W R, (1965), *Cognition and thought: An information processing approach,* John Wiley & Sons, New York, NY.

Rosch E, Mervis C B, Gray W, Johnson D, and Boyes-Braem P, (1976), Basic objects in natural categories, *Cognitive Psychology*, 7, 573-605.

Rosson M B, and Carroll J M, (1995), Narrowing the specification-implementation gap in scenario-based design, JM Carroll, Ed. *Scenario-based design: Envisioning work and technology in system development,* John Wiley & Sons, New York, NY, 247-278.

Rosson M B, Maass S, and Kellogg W A, (1988), The designer as user: Building requirements for design tools from design practice, *Communications of the ACM,* 31, 1288-1298.

Schön D A, (1967), *Technology and change: The new Heraclitus,* Pergamon Press, New York, NY.

Schön D A, (1983), *The reflective practitioner: How professionals think in action,* Basic Books, New York, NY.

Schön D A, (1987), *Educating the reflective practitioner*, Jossey-Bass, San Francisco, CA.

Scriven M, (1967), The methodology of evaluation, R Tyler, R Gagne, and M Scriven, Eds. *Perspectives of curriculum evaluation,* Rand McNally, Chicago, 39-83.

Tanaka T, (1994), *Designing scenarios for the near future: To enhance planning and development skills*, Design Research Institute, Tanioka Education Foundation, Tokyo, Japan.

Tversky A, and Kahneman D, (1974), Judgements under uncertainty: Heuristics and biases, *Science,* 185, 1124-1131.

Weidenhaupt K, Pohl K, Jarke M, and Haumer P, (1998), Scenarios in system development: Current practice, *IEEE Software,* 15(2), 34-45.

Wertheimer M, (1938), Laws of organization in perceptual forms, WD Ellis, Ed. *A Sourcebook of Gestalt Psychology*, Paul, Trench, Trubner, London.

Wexelblat A, (1987), *Report on scenario technology,* Technical Report STP-139-87, MCC, Austin, TX.

Wirfs-Brock R, (1995), Designing objects and their interactions: A brief look at responsibility-driven design, JM Carroll, Ed. *Scenario-based design: Envisioning work and technology in system development,* John Wiley & Sons, New York, NY, 337-360.

Young R M, and Barnard P B, (1987), The use of scenarios in human-computer interaction research; Turbocharging the tortoise of cumulative science, *Proceedings of CHI+GI'87: Conference on human factors in computing systems and graphics interface* (Toronto, Canada, April 5-9), ACM, NewYork, NY, 291-296.

Chapter 12

The Mind in Design: A Conceptual Framework for Cognition in Design Education

Rivka Oxman
Faculty of Architecture and Town Planning
Technion
Haifa, Israel

Abstract

Irrespective of the specific design domain, traditional educational models in design education are based upon the replication of professional task performance. The measure of learning is generally equated with the evaluation of the product of designing rather than on what might be considered a learning increment. *The cognitive properties of design learning have never been the subject of design education.* As a consequence, there presently exists a lack of educational theories of learning which function as an underpinning of design education.

1 Introduction

In the last decade, a considerable body of design research has begun to increase our understanding of the cognitive properties of design, and has provided new directions for the development of design education. In order to enable the construction of a general theoretical foundation, we argue the need to redefine the *learning task* in design education from an orientation to the production of design artifacts, to a cognition-based approach. In such an approach we believe that the design student can be *educated in designerly thinking*. Furthermore, a cognition-based approach can be structured in levels commensurate with stages of design learning.

Cognitive approaches in design and the role of knowledge and representations as a cognitive design thinking tool were discussed by various design researchers (Goel, 96; Schön et al. 92; Eastman 70; Akin 86). Most of these studies are related directly to design and design thinking rather then to the *learning task* in design learning and design education. In this paper we consider the role of cognitive content of design and design thinking as a basis for developing an educational approach. In order to do so we also consider *learning* itself as a cognitive function of the mind, and learning in design as a unique case.

In our research we have focused on cognitive issues regarding knowledge and representation. However, our assumption is that cognitive science in the provision of an understanding of how the mind works in such phenomena as knowledge acquisition, creativity, learning and development is a significant body of knowledge with great relevance for *design education*. Bringing these issues in the context of design learning have raised the following questions: how can design knowledge be acquired and learned? How can internal representations of knowledge be learned and taught? How can internal representational capabilities of the learner be increased? And how can the learning of knowledge and representation be evaluated?

In this paper, a conceptual framework is proposed in which the explicit learning of design knowledge structures and related cognitive strategies are the main objectives for design education and design learning. *Thus the cognitive content of design thinking becomes the subject of the educational program.* This approach is based upon the student's exploration and formulation of knowledge structures and reasoning in design, which are related to the particular design domain. We first propose a conceptual framework of the cognitive content of design thinking, which can function as a basis for design education. This includes the following four components: knowledge, representation, reasoning and cognitive models. We then present one possible framework in educational theory through which the cognitive content of design can be developed as an educational format. A series of learning exercises are described and their results as a medium of design education are evaluated. Finally we consider the problem of the evaluation of design learning, and the possibilities which are inherent within this approach. In conclusion, we discuss the potential for programs of cognition-based design education.

2 Tradition and change in design education

2.1 Design education

Among the objectives of design education is the establishment of competence to perform a body of cognitively high-level behaviors which we globally refer to as design (Finke, Ward and Smith 92). Design education facilitates our learning of these behaviors and our ability to enter into the professional realm of a specific design domain at a requisite level of skill and knowledge. Among other things, design education is that medium which enables the transition from the complete novice status of the beginner to the well-initiated status of the graduate designer. Thus the complex body of concepts, diverse forms of knowledge, instrumental skill, judgmental ability and professional norms and practices which must be transferred in the educational process is enormous. In order to achieve the complex mission of educating the designer, various cultures have developed theoretical positions for teaching design. These positions are deeply embedded in professional traditions and norms and are of historically long duration.

Modern design education inherited and transformed educational paradigms as part of the dramatic changes, which took place in the culture of design during the first decades of this Century. Modernist approaches to design education significantly changed the dominant educational paradigm while maintaining certain of its components. Essentially developing out of new cultural conditions as well as new educational theories, the modernist paradigm shift in design education has proved so fruitful an educational complement to traditional approaches, that it has been maintained even until today.

Despite its utility and durability, the modernist educational paradigm has begun to give way to new educational insights and theoretical developments. To a great extent this recent phenomenon derives form the two factors of the introduction of new educational approaches which have relevance for design education as well as the growing understanding of the cognitive properties of design through an expanding body of research. As a result of this, we are now witnessing a concomitant interest in design education as an area in which traditional educational approaches can potentially be expanded by this new cognition-based orientation.

2.2 Three paradigms of design education

2.2.1 The atelier system and the design problem

The first major paradigm of design education has rich roots in professional tradition. The *atelier*, the precursor of today's *design studio,* was a medium for conveying the content of design processes through supervised practice in the solution of design problems. Cuff (91) presents a well-accepted description of the centrality of the studio in architectural education. She defines the sources for patterns of studio education as derived from the famous French design institute, the Ecole des Beaux Arts. Certain of these patterns include the setting of problems as the initiation of the educational process. Setting the studio as a simulation of the professional environment. Setting the content of studio methodology as a series of well-formulated steps of design process, such as the *esquisse* stage, and the graphic formulation of the conceptual design. The relationship with the studio master as a tutorial relationship were based upon design documents, demonstration as a medium of communication, and the jury system was the forum for evaluation of the final product of design.

Established in the Eighteenth Century as a vehicle for *problem-oriented education* through the academic simulation of professional design practice, the design studio and problem-oriented education have become the dominant educational vehicle for diverse design fields such as architecture, industrial design and graphic design (Coyne and Rosenman 90). In the studio and in problem-oriented education the student acquires design knowledge through the guided solution of holistic design problems. Traditionally, there are many other attributes associated with the studio, for example, it acts as a venue for the absorption of norms as well as a medium for introducing younger novice students to the lore of successful design performance and the acquisition of skills. The problem sequence in the curriculum is also intended to convey a desired range of typological knowledge.

From an educational point of view the problem orientation of the design studio also functioned as an example of experiential learning by doing. In Dewey's formulation of experiential learning, knowledge and skills may be readily acquired and enhanced "by doing", that is, through accomplishing of actual work that relates to the goals of the acquisition of professional competence and of personal development. The approach enhances the effectiveness of formal education by placing emphasis on relevant competence, working knowledge and information and the development of skill in the actual doing of design. Piaget also contributed to our understanding of problem-oriented design education as a vehicle for knowledge development through stages of understanding that come as a result of interacting with, and acting upon, concrete problems. Thus the sequence of problems, as well as the social venue of the studio, contributes to the development of requisite knowledge and skills.

Despite its omnipresence as a vehicle of design education, problem orientation and the studio have a great range of educational variants and certain inherent deficiencies which qualify its educational effectiveness (Schon 85).

2.2.2 From studio to laboratory: formal exercises Vs the problem-oriented approach

The second major paradigm of design education derives from the great body of design educational experimentation of the Twentieth Century which may be said to have chrystalised in the pedagogical approaches of the foundation courses (Vorkurs) of the Bauhaus and the Russian State Higher Art and Technical Studios (VKHUTEMAS). In contrast to the atelier approach, these courses were generally non-project oriented sets of design exercises which were intended to develop an understanding of two and three-dimensional formal principles as a basis for designing. This was accomplished through structured exercises, which involved the exploration and explication of the abstract contents and characteristics of form. The exercises were primarily visual in orientation and intended to develop the "visual literacy" of the designer.

Thus the main thrust of modern approaches to design education such as the Bauhaus, and later at the HfG Ulm, and elsewhere has been not only to deal with the objects of design, but to enrich the understanding of the general formal content of these objects, etc. It introduced an orientation to design education as the derivation of knowledge through the exploration of general design principles - and not on the design process itself. As developed in the Twentieth Century this approach to education has been rich and diverse, and has generally been incorporated into the atelier system. Though I am unable to develop these points here in greater detail, it is important to state that this educational approach was rich in its psychological and educational theoretical sources including theories of perception, developmental psychology as well as educational theories.

2.2.3 Towards a conceptual framework in cognition

Despite these theoretical changes, and the important focus on visual and formal content, the educational focus still remains on the representation of the design object, rather than on an explicit articulation of the knowledge to be derived from the educational process. There is a need for the development of an approach to knowledge acquisition, which may become more educationally meaningful than the design object itself. *The third paradigm is the education of designerly thought processes in design reasoning and design strategies.* This includes how designers think in various characteristic design domain problem types, in reasoning, in exploiting graphical representations, in employing multiple forms of reasoning in design strategies. Such an approach also stresses the understanding of the character and role of domain knowledge, structures of knowledge, and the exploitation of knowledge in the processes of design.

By contrast to the explicitness of knowledge which must become part of the design educational process, the design studio today is still characterized by the faults of product orientation, creative design as a black box, and the pedagogical distance of the tutor. In all of its institutions there is generally a lack of explicit definition of the requisite knowledge of design, and a neglect of attention to thinking in design as legitimate pedagogical content. Although there may be an integration of design concepts, formal skills and knowledge which are "learned be practice" in the studio, the explicit learning of the cognitive content of design is ignored and left to be gained implicitly through experience. In the studio there are no effective methods of assessment to measure the developmental processes of learning on the part of the student in his education.

In the following sections we consider how cognitive science can provide theoretical frameworks relevant to design education. As an initial step towards the third paradigm we propose *a general theoretical framework for categories of cognitive phenomena in design* which appear to be of central and formative importance to both design and design education. Secondly, if we wish to impact upon the traditional educational situation, we must first *identify approaches to learning and educational methods, which appear relevant to the unique cognitive aspects of design.* Finally, we must find a means of introducing these new methods within the ongoing frameworks of conventional design education.

3 The cognitive content of design as educational content

Knowing and learning are among the foundations of education. We have seen that in traditional educational approaches the cognitive phenomena of knowing and learning are subsidiary, or implicit, factors in the development of educational programs. Our assumption is that cognitive science in the provision of an understanding of how the mind works in such phenomena as knowledge acquisition, creativity, learning and development is a significant body of knowledge with great relevance for design education.

What then are the implications of these assumptions and how can we establish a theoretical foundation for a cognition-based approach to design education? In the following sections we discuss topics in the areas of *cognitive modeling, knowledge and the formation of knowledge structures, representation, and reasoning*. We propose that these four subject areas be of central importance to design as well as to design education. Furthermore, they are among fields in which work has already been accomplished in cognition and in design research. There are many well-known works which can be applied to design thinking and exploited in design education: Mental imagery and visual reasoning (Kosslyn 75); Analogical reasoning (Holyoak and Thagard 95); Reminding (Schank and Abelson 77); thinking and creativity (Smith and Osherson 95); metaphorical reasoning (Lakoff 94).

3.1 Cognitive modeling

Diverse intellectual disciplines which study human cognition attempt to develop models of cognitive performance in various behaviors. *Cognitive modeling* is a root concept in the study of mind and one, which constitutes a common denominator among the diverse fields of the cognitive sciences. In psychology, for example, modeling can be employed to formulate such phenomena relevant to design as the way that people form and apply concepts, the speed of thinking with mental images and the performance of solving problems using analogies. In general, within cognitive psychology, cognitive modeling involves the formulation of information processing models, which can be evaluated with respect to a body of experimental data and empirical evidence.

In cognitive science, cognitive models are abstract models expressed as symbol systems and a set of processes, which manipulate them. Cognitive modeling forms a common platform for research in artificial intelligence which has provided a basis for theoretical understanding of various domains. There has also been a strong interface between cognitive modeling and computational systems in AI in design. This is of particular relevance in areas of computational modeling and systems development, which are dependent upon cognitive content.

In design, cognitive models of design thinking, which can be empirically verified, constitute one important platform for cognitive psychology in design research. For example, concepts and methods which have been developed in cognitive psychology have been applied to design in, among others, the work of Akin (Akin 86), and the work of Schon (Schon and Wiggins 92) particularly on visual thinking through interaction with the conceptual processes of design thinking. Cognitive models of design constitute a general approach for modeling cognitive process in design as global *processes* of thought and reasoning. For example, cognitive modeling in design has been employed in knowledge-based systems (Coyne et. al. 90), rule-based systems (Oxman 90) and case-based systems (Kolodner 93) in which such cognitive phenomena as classes of knowledge, domain expertise, and experiential knowledge can be symbolically formulated.

Cognitive modeling is the symbolic representation of phenomena in design. We can also employ cognitive modeling as a didactic medium which enables the student to better comprehend the richness and complexity, as well as the formality, of thought in design. Based upon design protocols, or self-protocols, the student acquires knowledge about design thinking by formulating models of the reasoning processes which underlay his and his colleagues own design. Thus cognitive modeling of design thus offers a research-oriented approach to design education.

Through our past work in the research of design processes and on the basis of our understanding of cognitive phenomena which are relevant to design there appear to be many subject areas in design which are amenable to teaching through modeling. However, we believe that there are three seminal areas of cognitive studies, which can be very profitably introduced as design educational content. These are:

- *Representation*, including the various symbolic representation of designs, and their manipulation during design, and particularly visual representations;

- *Reasoning*, including the symbolic representation and formulation of thought processes in cognitive phenomena such as reasoning from past design solutions (cases) or reasoning with analogy and metaphor, reasoning with visual images;

- *Knowledge*, including the representation and formulation of the various classes of knowledge in design such as typological knowledge, and the cognitive building of knowledge into larger structures.

3.2 Representation, reasoning and knowledge structures

There are several ways to apply the study of mind to design education. The first is to adapt subject areas from cognitive studies to design subjects. The second approach is to identify the different mental functions of the mind and to exploit their relevant theoretical sources and models in design applications. We have adopted the representational approach and propose that the general categories of representation, reasoning, knowledge and models are relevant to design education. Certain of the broad range of sources, which we have employed, include the fields of cognitive psychology, AI and AI in Design, and Computational Modeling. Among relevant sources are: concepts (Minsky 75; Schank and Abelson 77), semantic networks, analogies (Holyoak and Thagard 95), cases (Kolodner 93), metaphors (Lakoff 94), images (Kosslyn 94) and connections (McClelland and Rumelhart 81). These can be further expanded to include the full range of cognitive phenomena in design. We briefly define these four categories and discuss their relevance and their potential for design education.

3.2.1 Representation

Representation is a key concept in cognitive sciences and one, which has profound significance for design. An accessible way to understand what theories of mental representation and computation have contributed to the understanding of mind is provided by Thagard (96). He views thinking as representational structures in mind which are operated upon by (computational) procedures. Mental representations go beyond the descriptive power of models. They constitute rigorous formulations of the symbolic content which in combination with certain operation, or procedural acts, can simulate and explain human thought processes. Thagard reviews six important approaches to mental representations: logic, rules, concepts, images, analogies and connections. These approaches to representation have been applied in design research and have been reviewed extensively in the literature (Coyne et al. 90).

A second meaning of the term refers explicitly to representation as the symbolic graphical representation of designs as a "language of design". The learning of the conventions of these graphical languages of representation is one of the central objectives of design education. How we exploit the intrinsic characteristics of the different classes of representation, from diagram to sketch, and how and why we manipulate them in design and exploit them in reasoning about and in developing designs are further focal issues which must be addressed in design education.

3.2.2 Reasoning

Thinking can be understood and modeled in terms of representational structures in the mind and procedures that operate on them. The exploitation of representational schemes and operations as a means to formulate models of reasoning in design thinking has been theoretically and experimentally valuable during the last decade (Gero et al. 90). In our work, for example, case based reasoning (Kolodner 94) has provided a powerful framework to formulate prior knowledge in design, the conceptual knowledge of design precedents (Oxman 94; Oxman 96) as well as the processes which exploit this knowledge in reasoning. Such approaches to reasoning *employ formal descriptions in modeling reasoning processes, which are proposed as characteristic of central modes of reasoning in design.* Given the important body of knowledge available from the cognitive sciences, we are now capable of defining and modeling certain of these seminal processes of reasoning in design.

In addition to other modes of reasoning in design, an important sub-class of reasoning processes, which are most relevant to reasoning in visual-perceptual processes in design, is *visual reasoning with mental images and visual representations.* Representation here may refer to such content as the formal structure of the external visual representation which can be mapped into an internal representation, and which enables the process of reasoning. In our work on visual reasoning, the interaction with the visual representation in a conceptual-perceptual process is one of the phenomena which underlies *visual design reasoning* (Schon and Wiggins 92).

Current researches are providing a theoretical background for the study of visual reasoning, its relationship to formal mental representations, as well as the interaction with the visual content and conceptual structures in design drawings (Oxman 96). Such work on the modeling of characteristic design reasoning processes and on the nature of visual reasoning through the interaction with the various classes of the graphical representations of designs must become a central subject of design education.

3.2.3 Knowledge and knowledge structures

The acquisition of knowledge is one of the fundamental objectives of education. In design, the types of knowledge are highly diverse and domain sensitive. Both the classification of relevant knowledge and the development of media for the transfer of knowledge are among the central functions of education. Beyond the domain definition of specific relevant knowledge, we can designate certain general and high level forms of knowledge as significant in design. Generic knowledge is an example of one such high level form of knowledge; expertise is another.

When particular components of knowledge are structured into larger constellations of knowledge we can refer to these as *knowledge structures.* Knowledge that the designer acquires about design is not just an accumulating information, and its structuring into knowledge structures is one of the important knowledge-related cognitive phenomena in design. Knowledge structures are one way to capture information about how people comprehend information, store, remember and apply knowledge (Galambos et. al. 86).

Significant work in AI has attempted to model design knowledge by employing representational schema in order to identify and explicate the internal mental structures, representations, strategies and scripts of knowledge acquired by designers in their work. The form and the content of these information models represent the conceptual structures of certain types of design knowledge and capture the various cognitive knowledge types.

Among such work are efforts to define conceptual structures of design (Goel et. al. 96; Gero et al. 90; Oxman 94).

3.2.4 Cognitive models: knowledge structures and their associated processes in design thinking

The study of design thinking has gained in understanding the nature of knowledge from work that has been done with various representational schema and formalisms. In a general sense, *representational schema* in design are formal constructs for capturing, acquiring and representing types of knowledge structures in design. *Reasoning processes*, which operate, on a schema can be considered as cognitive process models of design thinking. Detailed schema for a specific domain or for a specific task are defined as *knowledge structures* (Abelson and Black 86)

Experiments with representational schema and formalization of knowledge structures have not only led to a better understanding of design but can also contribute to the development of our conceptual structure for design education. Formulating the content of domain knowledge by representational formalism such as semantic nets will be demonstrated to provide a powerful medium for explication of the structure of knowledge in design. Here again, we have exploited knowledge representation as a didactic means to teach about the specific knowledge structures in a particular design problem framework.

Any theory of knowledge must make a commitment to particular content schemas. Certain processes depend upon the content of knowledge. In our work we have employed knowledge structures, e.g. design case knowledge, to model and demonstrate cognitive design phenomenon in design thinking. For example, we have presented schemas for the representation of typological knowledge and a procedural approach to generic design as a formalized reasoning process. In one case, the representational schema and the reasoning mechanism were incorporated in an expert system (Oxman 90) which helped us to demonstrate the top-down process of 'typological refinement' as a characteristic thinking process in design. We have also attempted to demonstrate creative thinking with typological knowledge. In this work we demonstrated that a specific formal processing of typological knowledge (in this case, a chair design) exhibits a phenomenon of creativity in design thinking which we have termed as a 'typological emergence' (Oxman 99) (see Figure 1).

Cognitive models are thus complex formulations of representations, reasoning and knowledge, which can model and explain complex, extended and hybrid cognitive phenomena in design. Beyond simpler characteristic reasoning processes, cognitive models can be employed to model more complex and compound processes such as design strategies. Understanding and modeling such phenomena may be considered an advanced educational objective.

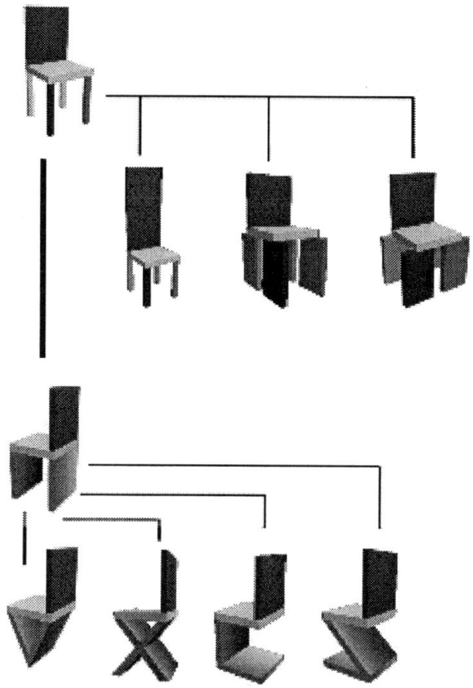

Figure 1: Typological emergence

4 Education approaches relevant to design education

In the following chapter we present certain educational and learning theories which appear to be relevant to design education. We will discuss the implications for design learning of these various educational frameworks, and what kinds of changes in learner understanding they imply.

4.1 The significance of "learning by doing" as a general educational approach

We have proposed that *learning through the structuring and manipulation of knowledge* in design may be considered a significant educational objective in design education. We believe that the cognitive attributes of design cognition and learning can become the content of design education. In order to develop an educational framework for design education, we must first attempt *certain basic definitions regarding what one learns when learning to do design*. We propose the following as preliminary propositions:

1. Learning in design is the acquisition of the cognitive ability to represent design knowledge, to acquire basic schema in design thinking, to understand knowledge structures such as generic and typological design, adaptive design, analogical and metaphorical concepts and images.

2. Learning in design is the acquisition of the cognitive ability to manipulate the representations of design knowledge, to understand knowledge structures and their associated manipulation strategies such as refinement and adaptation, concept formation, etc.

3. Given the first two hypotheses regarding *the content of design learning*, we can then propose that design education is the processes and media, which contribute to design learning. We suggest that an effective medium of design education be in modeling the representation of design thinking. Through modeling of knowledge structures and strategies the student gradually develops a conceptual understanding of design. We have found that such *a research-oriented approach to design education* can be a lucid medium of design education. Furthermore, *the three paradigms of design education may be seen as a complementary set.*

In the following sections we define relevant theoretical sources and identify relevant cognitive approaches to learning and educational methods, which appear relevant to the cognitive aspects of design thinking. We first demonstrate two examples of cognitive theories, one of development and one of creativity, which appear to us to be of relevance to design learning, and with which we have experimented in our design educational programs. Following this we briefly introduce Papert's theory of "Constructionism" (Harel and Papert 91) which has provided us with a general educational framework.

4.2 Explication of knowledge structures as a form of learning

The theory of representation and re-description (RR) (Karmillof-Smith 95) is relevant to design learning and the explication of the interaction between the visual and the conceptual content of design. This model is important, since it refers to learning as the succession of representations which becomes progressively more manipulable and flexible for the emergence of conscious access to knowledge structures. According to this theory the human means to gain knowledge is for the mind to exploit acquired information that it has already stored as general schema by re-describing its representations. For example, in one example in the cognitive development of the child, the understanding of the concept and form, "man" provides a schema for the re-description of the concept in a graphical derivation such as "funny man". In the process of re-representation implicit knowledge of the schema becomes externalized in the sequence of representations. The process of representation and re-description suggested by Karmilloff-Smith is relevant to design learning, since it involves *conscious construction and exploration of the content of the schema of knowledge that is employed in design thinking.* We can summarize that learning according to this theory of development is a process of the *acquisition of knowledge structures, which are related to schema differentiation.*

4.3 Explication of cognitive strategies exploiting knowledge structures

Another important principle for design learning, is *the interaction between visual and conceptual content in global strategies of design thinking.* An example of the explication of this interaction in characteristic strategies in design thinking can be found in the theory of Creative Cognition (Finke, et al. 92).

Formal structures of global processes can be described, or characterized, in terms of models explaining how conceptual knowledge is employed. Global strategies in the cognitive sense refer to the larger structures of cognitive phenomena, which characterize the exploitation of knowledge in thought. That is, global strategies in design such as typological or analogical thinking can be modeled as complex graphical schema. These models of design strategies build on *two principles of design representation, which can be defined: the fundamental duality of graphical and conceptual content, and the creation of knowledge structures as basic categories of design knowledge.*

In global strategies of creative design the strategy is less formally based on generic knowledge and developmental processes, and is more exploratory. This different class of strategy illustrates the distinction between "processes used in the generation of cognitive structures and those exploring the creative implications of those structures"(Finke et al. 92). In creative cognition, cognitive structures that are generated have emergent properties that can be explored, where some of the properties could not have been anticipated in advance. In this way one might generate radically new and unexpected ideas.

The theory of creative cognition thus provides a basis for addressing the significance of knowledge structures to creativity, as well as a theory developed within a framework of the interaction between visual and conceptual content.

4.4 Learning through construction

This approach is motivated by Constructionism (Harel and Papert 91) as a theory of learning which emphasizes developing an epistemology of knowledge through construction. The constructional form provides a representation of the structure of knowledge, which the student acquires through constructing. Learning may be considered a process of knowledge acquisition and development in which the knowledge, or a representation of the knowledge, is physically constructed. There are many other educational attributes to the Constructionist approach including the communicative function of media (the student as teacher) and the public nature of the process of construction (the student's work as part of a larger group effort).

Within this general pedagogical approach the goal of design education is defined as the acquisition of design knowledge through constructing the explication of schema, knowledge structures, and global strategies in design thinking. Through constructing representations of design thinking the student gradually becomes richer in his ability to think in designerly ways. This contributes to an understanding of the cognitive processes, which are characteristics of design, or as Papert has stated, this form of education contributes "to knowing rather than to knowledge".

5 Mind in design: a conceptual framework for cognition in design education

5.1 The model: representation, construction and implementation of knowledge structures

5.1.1 Representation

In recent work we have employed the semantic network as a representational formalism which we have used to represent design knowledge structures and to model reasoning processes. We have also employed, as in Figure 1, the network structure as a basis for modeling design thinking and design development as represented by the evolutionary structure of the design image. These techniques are proposed as examples of representational media, which can be effectively exploited in cognitive modeling in design.

From a content point of view, we have developed taxonomy of basic elements in cognitive processes. These include the characteristic interaction between design strategies and form generation. This taxonomic code enables the modeling of cognitive structures and strategies. These are modeled as network structures in the form of nodes and linkages. The network structures explicate, and enable comparison between, various classes of design knowledge and cognitive strategies in design.

The elements of the network are based upon formalism previously developed for representing design thinking. The representational formalism termed, ICF (Issue-Concept-Form) (Oxman 94; Oxman 99) addresses problems of representation of knowledge in design. ICF represents chunks of knowledge of designs, provides explicit linkages between the issues of a design problem (I), a particular solution concept (C) and a related form description (F). It has proved powerful in representing the ideational content in design which links graphical and textural content, and it has been employed to represent typological knowledge, generic reasoning, refinement, adaptation and other cognitive phenomena in design. The formalism has been expanded in order to include analogy (A) and metaphor (M) as resources for ideation and to support design exploration processes, which exploit these cognitive mechanisms. Examples are provided in the following sections.

5.1.2 Construction

According to Papert, "constructionism, shares constructivism's connotation of learning as building knowledge structures". We are motivated by the general constructionist approach that learning through construction can be a medium for building knowledge structures in the mind of the student. By constructing models of these structures, the learner acquires knowledge of the cognition of designerly thinking. That is, he learns the cognitive processes of thinking design as well as the cognitive structures of knowledge, which are employed in designing. Further with respect to constructionsim, the learner is consciously engaged in constructing a public entity. This work benefits from the public format in which the models constructed have a communicative, as well as learning, value.

5.1.3 Computer implementation

In our work we have elected to exploit the *computer laboratory as a design learning environment.* The venue is of particular significance in its own emergent properties in design education. We would like to propose two important justifications for the computer lab as a *new venue of design education.* Firstly, we believe that the lab is becoming the locus for the synthesis of the three educational paradigms. Our students have readily moved between project modeling, design principles, and cognitive modeling. They appear to have benefited educationally from the interaction and interchange between the three modes. Secondly, we believe that the computer laboratory provides a complementary educational venue to the studio and one, which is strongly and articulately theoretical in nature.

As design educators we use computers as educational tools in simulating cognitive processes; explicating knowledge structures and even gathering their content as potentially useful material for the designer, such as computer-based libraries, visual precedents, etc. This way, the computational system supports both the learning process as well as the potential re-uses of this knowledge. In this case the computer is the medium for both the construction of knowledge and the modeling and experimental testing of the cognitive content of design thinking (Oxman et. al. 97).

In the following section, specific examples of student's exercises are presented and the teaching program is evaluated as a contribution to a cognitive-based design education.

6 A teaching program in the representation, construction and implementation of knowledge structures

We tested our hypothesis with a group of graduate students from the industrial design and architectural programs at the Technion, Israel Institute of Technology. These were primarily upper level undergraduates and second-degree students, who had both computational ability and some background in the theoretical design context of the exercises.

Our work on design education is situated in the context of specific modeling tasks. In response to these tasks design thinking is formulated as conceptual structures and strategies by means of a network representational formalism. These structures formulate semantic nets of the conceptual space as well as the content of their related key design ideas. The representations may be considered to describe significant solution paths within problem genres of various design domains.

It is difficult to develop a self-awareness of design thinking through conventional design activity and introspection. However, virtually all of the students developed proficiency with this methodological foundation of design thinking. We will describe such learning processes in the following examples. The diagrams illustrated below are schematic representations of actual student work.

6.1 Representing and modeling generic and typological design

This exercise appears to have provided an efficient and articulate medium for understanding and acquiring skill in generic thinking. Students were given a brief theoretical introduction to generic and typological design. Generic design was presented as a form of knowledge structure that connotes the body of prior knowledge, which enables the designer to evoke generic representations, or to extract generic schema from specific images. It includes both the derivation of generic representational schema as well as the knowledge of exploiting these schema (Oxman and Oxman 92).

Students were asked to represent typological knowledge as a set of generic representations which are associated with specific problem types, and to organize the variables of the type in a hierarchical order of which the highest level is that of the schematically represented class description.

6.1.1 Representation and construction

A basic coding of thinking process in interacting with the representation includes an issue (derived from the problem), the resulting concept, and the resultant form (see Figure 2).

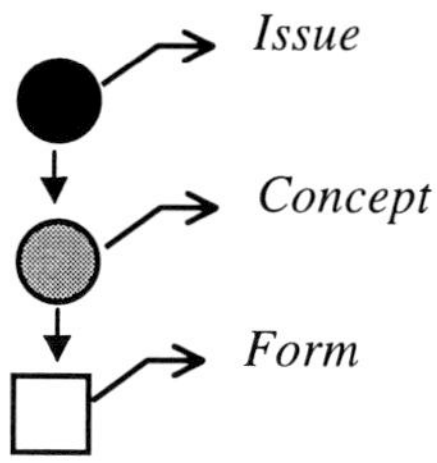

Figure 2: Simple cognitive process employing a problem specification, solution type and the resultant form

This diagram was exploited to model generic design and typological design as follows. In the first case, a representation of generic design is presented as the derivation of a solution class, which then results in the production of a "generic representation". This is a recognized phenomena in design which results in a schematic graphical representation (level 1) which is subsequently developed as a realized scheme (levels, 2, and 3 and so forth). Figure 3 illustrates this process of generic design.

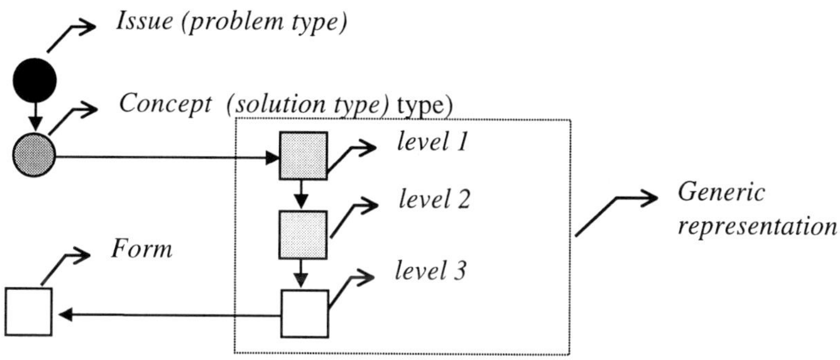

Figure 3: Generic representations in a refinement process

This general model of generic design was extended to accommodate the knowledge of a specific building type. Thus another global developmental strategy which builds upon generic thinking to exploit typological knowledge is commonly referred to as typological design. In Figure 4, typological knowledge is represented as a variation on generic design in which each refinement level of the generic representation initiates reasoning regarding issues and solution concepts which derive from knowledge of the building type, its sub-problems and sequence of significant design variables. This way, generic thinking is guided by typological knowledge, one well-known design phenomenon.

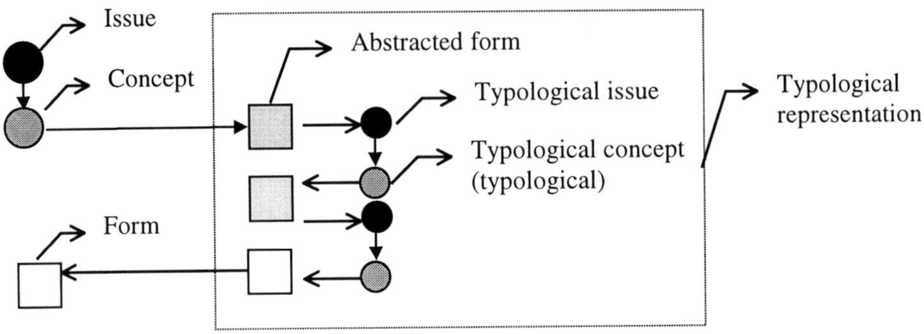

Figure 4: Typological representation in a refinement process

The construction and the final implementation of the generic representation of the student's work were dynamic and interactive. The user could interactively modify the form according to the typological variables and the generic structure of knowledge The implementation in this example was of a chair design which supports generic and dynamic back and forth particularization through navigation between hierarchical levels of abstraction. An interactive environment employing VRML (virtual reality modeling language) provided for design zooming which simulates the generic representational levels of a specific class of chair designs. In this way, the user can alternate between a schematic and particularized representation through the medium of zooming. Each of the abstraction levels may be elaborated by adding more specific details.

6.2 Representing and modeling adaptation through re-representation

This exercise has provided a significant medium for understanding the role of representations in adaptive design. In the theory of design adaptation a specific design is selected and, through adaptation, transformed into a new design. In comparison to generic design it is represented as a specified design representation. Adaptation processes can be defined as successive modifications through a series of representations, which are executed upon the original design. We have referred to this process as *re-representation* (Oxman 96), exploiting the term which has been applied to the cognitive phenomenon as well as to the cognitive capabilities which make these complex processes possible.

6.2.1 Representation and construction

Students were asked to identify and present these classes of representations as well as other sources with which designers can interact, and modify. This is an example of their coding of multiple representations. The representations may be derived through interacting with exploratory visual resources such as analogies and metaphors. Figure 5 presents derivations taken from multiple representations.

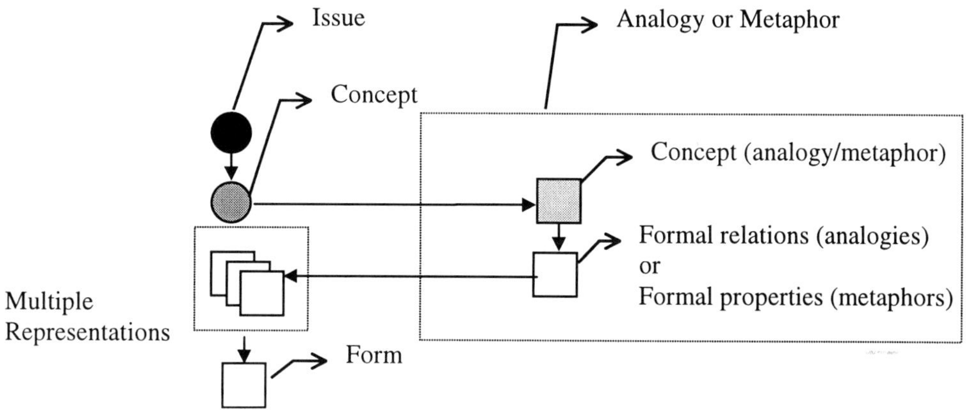

Figure 5: Representation of multiple representations and their mental resources

6.2.2 Computer implementation

It was suggested to the students that one way to support adaptive change be by providing a medium to interactively construct explicit multiple representation of sub-structures.

6.3 Representing and modeling visual-conceptual ideation in precedent-based design

Precedent-based design is accepted as one of the cognitive phenomena in design creativity and as a source of ideation. The generation of a conceptual basis for the inception of design is one of the most interesting of the cognitive phenomena of early design. The students were asked to consider series of issues such as how precedents should be represented, indexed, and organized in order to support the process of ideation. Rather than selecting specific precedents, it was the exploration of their conceptual structure, which was to be supported. Students were asked to construct a *resource for design ideation* in which the conceptual content of design precedents acts as a vocabulary of design ideas within the framework of a particular class of design problems. They were asked to employ a *concept vocabulary* of the resource base of selected precedents in the construction of a semantic network of ideas, which can be browsed by the designer as a cognitive resource for design ideation.

6.4 Representation and construction

With respect to exploratory thinking, the following student employed the same principles of graphical coding to distinguish exploration. The student employed the code to represent complex networks. Furthermore these networks introduced references to specific designs, or precedents. This is represented in Figure 6, which models associative reasoning and the exploitation of precedents in design. In another work, student presented complex hybrid processes including both exploratory and developmental sequences.

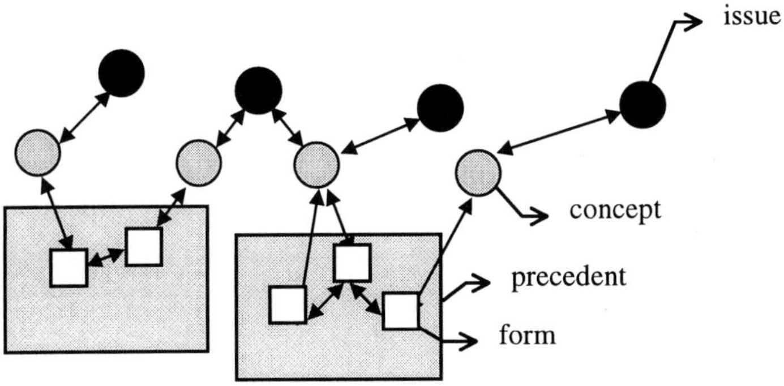

Figure 6: Representation of associative browsing among multiple precedents

6.4.1 Computer implementation

An implemented system of design precedents was developed (Oxman 00). Through the employment of this system (WEB-PAD) the student may browse to explore how a similar conceptual form may be realized in different designs. Conversely, by selecting specific precedents the user may explore other concepts, which were not expected and emerge as a result of the navigation process.

7 Problems of evaluation of learning experiences

As we have stated, in most conventional approaches to design evaluation, only the final product is evaluated and not the cognitive learning increment of the student. How can design learning be measured and evaluated?

Although we have not undertaken a rigorous form of analysis to demonstrate that subjects in our teaching program did improve their ability to design we have obtained some significant insights and qualitative results. The validity of the qualitative results of this experiment derives from the use of mixed methodologies of assessment such as formal interviews with participating students during and after completion of their projects as well as discussions, criticism, evaluation of student's work and observation of progress. Based on these assessments, we could make observations of changes that occurred in student's thinking about design, their growing skill to deal with the complexities of design thinking and the dynamic progress of their performance.

7.1 Evaluation of representation and construction

In the brief example below, we refer to *the development of concepts and knowledge structures in design learning* as one of the most significant qualitative results we obtained in our program. This is a significant category in measuring design learning.

The qualitative assessment of learning as the development of concepts and knowledge structures in design learning was measured in our experiment by an evaluation of modeling performance. We have traced the use of concepts according to their ability to draw inferences concerning their acquisition of design knowledge. We defined linkages and terms of the knowledge structures. For example, in employing conceptual linkages we assessed the collection and organization of knowledge through such indicators as the type of linkages, and the degree of development of the structure of knowledge represented. Such measures include the types and the number of conceptual linkages, employed. We illustrate evaluation in the following example from one of our students. In the assessment of the evolution of a representation in the work of one student, we briefly analyze the learning processes, which was observed during the interaction with the teacher. The project involved the representation of ideation processes in precedent-based design.

The first step of the student was to model an example of a design idea in a particular precedent. For example, an idea such as achieving phenomenological content in housing design can be achieved by use of traditional elements such as an inner court. The conceptual content and the form of this particular example was presented as a simple reasoning process, linking between issue: phenomenological content concept: use of traditional elements, and the form: inner court. This is illustrated below in Figure 7.

In dealing with two precedents which had the same formal solution, gradually a greater generality and a greater ability of generic understanding appeared in the use of the concept "inner-court". The need developed to employ a higher form of conceptual representation such as prototype and generic solution.

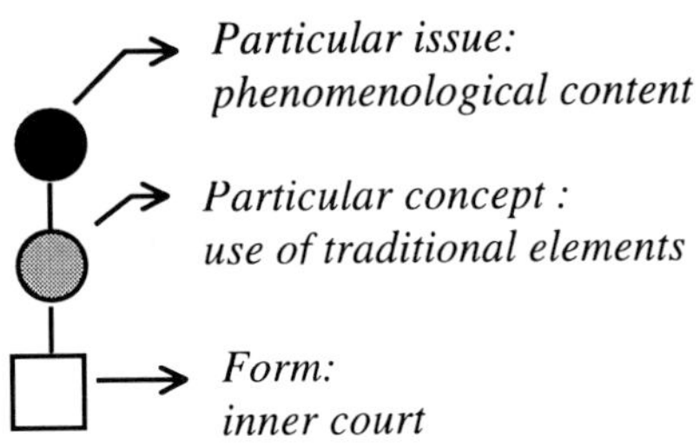

Figure 7: Development of concept (a): particular issue, concept and form solution in a design precedent

The second concept which developed in the student's working with the representation of ideas in concrete case studies is the need for generalization, or "association through an abstraction level" which is modeled by links between the two similar design solutions that employ the same conceptual solution.

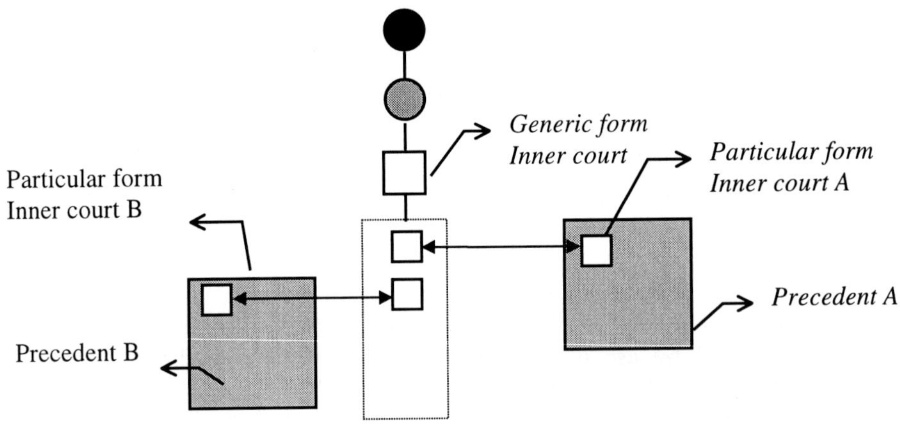

Figure 8: Development of concept (b): generic form and associative thinking

The third representation developed the notion of an abstract concept. We can see that the concept "use of traditional elements " can be also realized by the use of "niches" in the design. Following this, the student expanded the concept of "abstraction" through the use of an abstracted solution concept.

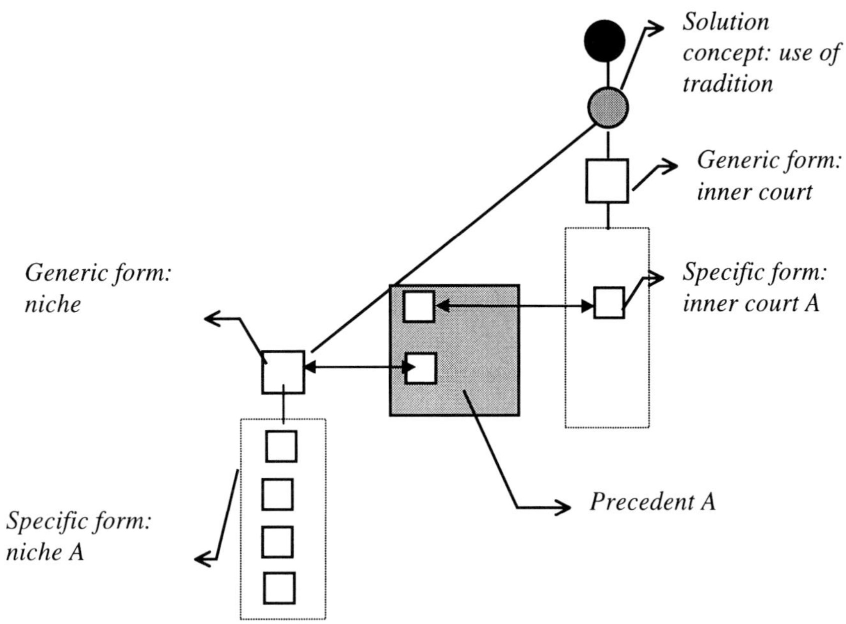

Figure 9: Development of concept (c): expanding the concept of abstraction.

In the next representation the student developed the concept of "exploration". By interpreting the links as browsing devices in which exploration has enabled the finding of new ideas in design precedents.

Having observed such phenomena as improved representational ability might not necessarily guarantee that the student is a better designer, but that he/she has definitely begun to be a designerly thinker. It is these kinds of evaluations based upon the observation of the growth of knowledge and understanding, which should become among the measures of design learning. This is not to say that we should abandon traditional design pedagogy and the actual production of design artifacts. But if we are design educators, we must find means to supplement traditional pedagogy by educating the designerly thinker as well as the maker of designs.

In the brief example above, we refer to the development of conceptual understanding, one of the most significant categories. In the following section we evaluate how the acquisition and the explication of knowledge structures through computational modeling and the use of the technology, have contributed to design learning.

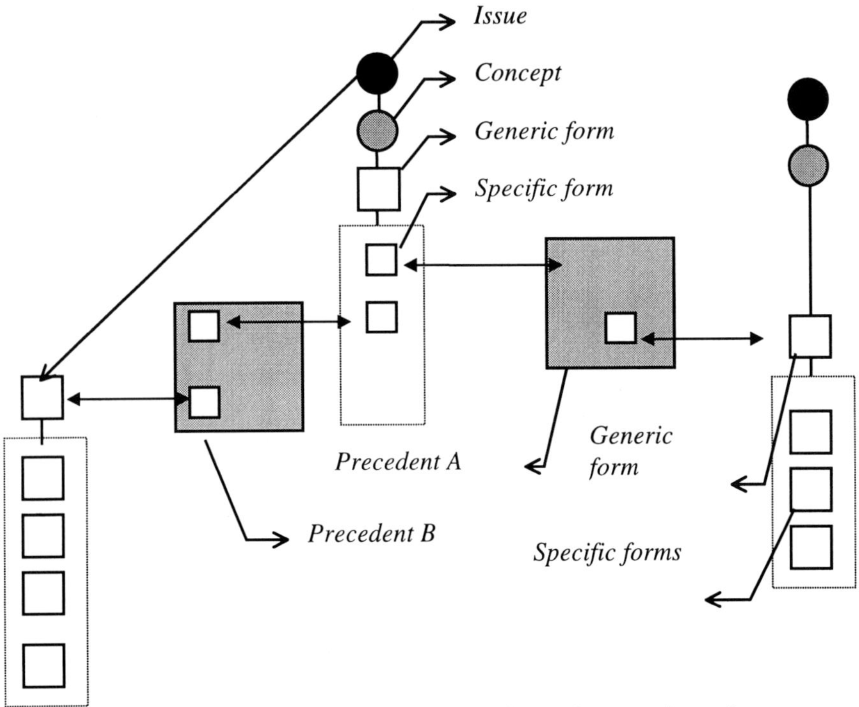

Figure 10: Development of concept (d): exploratory browsing

7.2 Evaluation of computer implementation

Venues of design education, the classrooms or the studios, are the spatial institutionalization of educational theories. The class-room may provide a spatial situation for directed analytical education in problem solving, while the studio is traditionally viewed as providing the creative ambiance of design making. How is it possible to spatialize the exploration of cognitive phenomena and the modeling of cognitive processes?

The practical techniques of this method focus on the development and the design of computational systems, which can represent the cognitive models. This has proven to be a powerful, stimulating and innovative educational medium for the acquisition of knowledge. This way, the computational system supports both the learning process and the potential re-use of this knowledge and the creative and learning increment of their work can be evaluated comparatively.

On the basis of our work we believe that this new venue has proven itself as a medium for design education. We should consider the significance of the venue and the medium from the point of view of their contributions to educational performance, and in comparison to conventional media and venues. The enhancement of collaborative, versus competitive, motivations of students is significant and may be enhanced in-group modeling. The ephemeral nature of the results of conventional educational activities can be compared to the potential value of knowledge permanence in the research-related character of this approach.

8 Summary and conclusions

It is now possible to demonstrate that the derivation of design knowledge through constructive processes, in itself, provides a medium for design learning. Though the observation of the process of construction we have identified such phenomena as *depth of understanding* of the task in the development of modeling skill, and the efficiency in modeling representation. A growing level of complexity in the representation indicates the acquisition of a deeper understanding of relevant concepts and the mastery of their interactions in design thinking; the sophistication of integrating and implementing concepts in complex structures may be interpreted as considered an indication of metacognitive insight. This is perhaps an indication of a high level of general knowledge regarding the interaction of knowledge in design strategies.

This approach transcends the educational logic of conventional venues of the classroom and the studio. It suggests that special design learning environments must be developed which can enhance and supplement formal education and foster personal development in design learning. As for the potential of future applications of this methodology, we believe that the resulting relationships between cognitive models of design, design domain knowledge and the incorporation of computational technology has theoretical and practical implications for design education in the broad spectrum of design domains.

9 Acknowledgments

This paper is a revised and extended version of, "Educating the designerly thinker," which appeared in a special issue of *Design Studies*, volume 20, no.2, March 1999 invited to be presented in a workshop by the editors of the special issue. The theoretical research and experimental implementations were supported by a DFG grant from the German Government and the Technion Fund for Encouragement of Research.

References

Akin O, (1986), *Psychology of Architectural Design*, Pion, London.

Coyne R D, Rosenman M A, Radford A D, and Gero J S, (1990), *Knowledge-based design systems*, Addison-Wesley, New York, NY.

Cuff D, (1991), *Architecture: the story of practice*, MIT Press, Cambridge, MA.

Eastman C, (1970), On the analysis of intuitive design processes, G T Moore Ed. *Emerging methods in environment design and planning*, MIT Press, Cambridge, MA.

Finke R A, Ward T B, and Smith S M, (1992), *Creative cognition, theory research and applications*, MIT Press, Cambridge, MA.

Galambos J A, Abelson R P, and Black J B, (1986), *Knowledge structures*, Lawrence Earlbaum, San Mateo, CA.

Gero J, Lee H, and Tham K, (1991), Behavior: A link between function and structure in design, *Proc. IFIPWG 5.2 Working conference on intelligent CAD*, Columbus, OH.

Goel A, Gomez A, Grue N, Murdock W, Recker M, Govindaraj T, (1996), Design explanation in interactive design environments, *Proc. fourth international conference on AI in design*, Palo Alto, CA.

Harel I, and Papert S, (1991), Software design as a learning environment, I, Harel and S Papert Eds. *Constructionism*, Ablex, Norwood, NJ.

Holyoak K J, and Thagard P, (1995), *Mental leaps: Analogy in creative thought*, MIT Press, Cambridge, MA.

Kosslyn S M, (1975), Information representation in visual images, *Cognitive Psychology, 7*, 341-370.

Kosslyn S M, (1994), *Image and brain: The resolution of imagery debate*, MIT Press, Cambridge, MA.

Karmiloff-Smith A, (1995), *Beyond Modularity*, MIT Press, Cambridge, MA.

Kolodner J, (1993), *Case-based Reasoning*, Lawrence Earlbaum and Assoc., Hillsdale, NJ.

Lakoff G, (1994), What is metaphor? J.A Barden and K.J Holyoak, Eds. *Advances in connectionist and neural computation theory, analogy, metaphor and reminding*, 3, 203-257, Ablex, Norwood NJ.

Marda N, (1997), Visual design thinking, *Stoa, European Association for Architectural Education*, 2, November.

Minsky M, (1975), A framework for representing knowledge, P.II.Winston, Ed. *The psychology of computer vision*, McGraw-Hill, NY, 211-277.

Oxman R E, and Oxman R M, (1992), Refinement and adaptation in design cognition, *Design Studies*, 13(2), 117-134.

Oxman R E, (1994), Precedents in design: A computational model for the organization of precedent knowledge, *Design Studies,* 15(2), 141-157.

Oxman R E, (1996), Design by re-representation: A model of visual reasoning in design, *Design Studies,* 18(4), 329-347.

Oxman R E, Sarid A, Bar Eli S, and Rotenshtriech R, (1997), A conceptual network for web representation of design knowledge, *CAAD Futures'97*, Kluwer.

Oxman R E, (1999), *Educating the designerly thinker*, C. Eastman, W. Newstetter and W. M. McCracken Eds. *Design Studies,* 20(2).

Papert S, (1991), Situated constructionism, I Harel S Papert, Eds. *Constructionism*, Eds. Ablex, Norwood, NJ

Shabo A, and Oxman R E, (2000), A WEB-based tool for the construction of visual case libraries, AID 2000 Fifth international conference on artificial intelligence in design, Worcester, MA.

Schank R C, and Abelson R P, (1977), *Scripts, plans, goals and understanding,* Hillsdale, Lawrence Erlbaum and Associates, N J.

Schön D A, (1985), *The design studio: an exploration of its traditions and potential,* RIBA Publications, London, UK.

Schön D, and Wiggins G, (1992), Kinds of seeing and their functions in designing, *Design Studies*, 13(2), 135-153.

Thagard P, (1996), *Mind: Introduction to cognitive science*, MIT Press, Cambridge, MA.

Chapter 13
Design Knowing & Learning: A Socially Mediated Activity

Louis L. Bucciarelli
School of Engineering
MIT
Cambridge, MA

Abstract

Why is teaching engineering design seen as different from teaching engineering mechanics, or instrumentation and controls, or thermodynamics? The author claims it is because design knowledge and knowing is very much context dependent in a way that the engineering sciences are not. Design knowledge is not simply information - easily transmitted to our students like a material substance - nor does it require instrumental *object-world* reasoning alone. Designing requires the negotiation of interests and proposals of different participants; hence the process is social and knowledge socially construed. The author illustrates how the context of the engineering classroom might be transformed to better accord with the context of engineering practice through the introduction of open-ended, design type exercises.

1 Introduction

Design, within the engineering education curriculum, has traditionally been viewed as a weak sister; while faculty can energize students with talk of creativity, and engage them in the act itself, the subject matter is considered "soft" and lacking in rigor when compared to courses in control theory, signals and systems, thermodynamics, computer science, or engineering mechanics. The analytic/synthetic distinction is alive and well in engineering education.

This chapter explores this dichotomy, not in order to deny the differences between science-based, theoretical knowledge and design thinking and knowing, but rather to understand these two forms better and how they relate to engineering practice. My thesis is that design thought, knowledge, and knowing is very much dependent upon context, and that context is social; for designing is generally a collaborative effort which engages individuals within (and outside) the firm. The methods of reduction and abstraction that appear to work so well in the shaping of engineering science subjects, fail to capture the collective nature of designing and design knowledge. (Bucciarelli and Schon 87) What can we say, then, about "design knowing" and about "design learning" from a perspective which views exchange and interaction among participants in design as essential to engineering design practice?

Among engineering faculty, the social character of engineering theories and practices does not garner much attention. True, engineers learn and exercise their theories and practices within particular social settings - the classroom, the lab, the design review - but generally theories and practices are talked about as if they stand independent of context.[1]

They are taken, like science, to be universal in nature, the same in Singapore as in London, Paris, or Boston. I want to explore the basis, in our approach to the education of engineers, of our almost total attention to instrumental theories and methods and neglect or discounting of context. I conjecture that our unease with the social derives from the way we think and speak about "knowledge" - as some sort of material substance. I argue that we need a more contextualized understanding of engineering knowledge and that this means we need to pay attention to the social character of engineering practice, in particular of engineering design.

I first describe a framework for understanding process from this broad perspective. I then summarize the main results of my research on design process, working from this perspective. I turn to the way we traditionally think of knowledge and show how that way is deficient; how it influences the way we teach and results in too narrow and instrumental a student experience. I then offer a way to transform the context of engineering education to focus explicitly on the knowledge and know-how required for engineering practice today.

2 An Ethnographic Perspective on Designing

What constitutes designing, a design task can vary depending on who you ask. Boundaries can be drawn close in, e.g., around an individual at a workstation, or further afield, encompassing all who have an influence on the design. Where one sets the boundaries, decides who is in, who is outside the process, depends upon one's interests and the questions one seeks to answer. Engineering design here is to be construed broadly. Design participants are taken to include all those who have an interest in and an effect upon the form and function of the product or system. Generally, an engineering design task is the

[1] By "engineering theories" I mean the science based, paradigmatic bodies of knowledge which underlie and frame our models, designs, plans and productions. By "practices" I mean the know-how of professionals at work in the world displayed as they design, e.g., know-how embodied in heuristics and rules of thumb, in interpreting constraints and specifications, codes and regulations, in experience with materials and methods of fabrication and the real time performance of products and systems.

province of a formally designated team within the firm. Officially team members or not, all of these individuals who weigh in on the task I consider participants in the design. These might include a supplier outside the firm as well as the project leader, inventory controller as well as project scientist, and of course people from marketing and manufacturing.

Regarding temporal boundaries: I take it that the process includes all those events and acts related to the object of design as participants move from antecedents in the form of ideas, last year's product, project specifications and the like, to a marketable, and possibly profitable product. Exact definitions of a starting time or ending time are often difficult to fix.

Boundaries around people and on time-as-duration are not the only delimiters that need to be established, the only dimensions that need to be set out. The "content" of design work also calls for definition. Here too my scope is more generous than that covered in most contemporary text books in engineering, texts which generally only deal with the instrumental rationality of design, which rely upon graphic representation - block diagrams, concept maps, matrices of interaction and tables of categories of design functions and properties - to define the design process, while listing rules and norms for carrying through that process (Hubka and Eder 88; Dixon 66); or those that lay out algorithms for optimization and other so called rational methods for designing. (Statnikov 99) My claim is that these offer too limited a perspective for understanding design as it is practiced and, hence, for fixing the depth and range of knowledge required for that practice. Too instrumental a perspective ignores or doesn't see as significant critical features of design process, namely those loosely structured occasions for negotiation among participants which are essential to concept formation and decision making; it misses matters of significance because the content of these occasions does not appear rational in the instrumental sense.

Generally speaking, I take a broader perspective when compared to others. My approach is ethnographic. In my study of process, I enquire about informal as well as formal organizational ties; about participants values, interests and vision as well as the information they access, the infrastructure they rely upon. I pay attention to language, the use of metaphor, and stories. The sophisticated skills and techniques of participants are given special attention. (Orr 90) Here I do not presume that tools and methods are used in the same way by different participants, even though they appear to be mechanical and rule bound, so the content of participants work needs to be understood. And, as I claim knowledge is constructed in action, context is all important. (Lave and Wenger 91)

The product of research in this mode is usually a story - a narrative intended to convey what's going on and what's significant in the lives of the "natives"- what constitutes their world, world view, and hence, how that might change, be changed, or react to change. The aim of ethnography is a good narrative. A good story is a truthful story; and like a scientific theory, can be put to the test, making sense to the participants in design themselves as well as one's faculty peers in engineering, in management. A good story relies upon metaphor as well as upon "the facts observed". Metaphor is necessary when we attempt to describe that which lies outside our common experience, our usual categories, and ways of representing. Yet integrity is essential; and here the discipline of anthropology has much to say.

Unlike a traditional technical theory, the descriptions ethnography produces are open to interpretations of different shades; different readers can interpret, experience ideas, with shades of differences; certainly different readers will be provoked differently. But this does not mean it is "unscientific"; the claim here is that a good story which has depth - a "thick description" in the words of Geertz (73) - in allowing for varied interpretations, avoids the over-simplification of a behaviorists model or cognitive scientist's mapping and as such is in closer touch with reality - however that is construed.

3 Design Process

3.1 Object Worlds

From this ethnographic perspective, the social nature of designing is evident at every turn. Yet instrumental rationality does have its place and a most important place at that. The paradigmatic sciences which rule the workings of products and systems have to be attended to and exploited if the design process is to succeed and culminate with the launching of a profit-making product.

In any contemporary design task, different instrumentalities are engaged by the different participants according to their different competencies. I say that different participants work within different "object worlds", the content of which is a matter of the individual's technical experience, training, resources and responsibilities.(Bucciarelli 94) Individuals are trained and educated and accumulate different experiences within different disciplines - structural engineering, instrumentation and controls, electrical systems, or mechanical systems, and the like - each with their own panoply of exemplars and paradigmatic sciences. Different object worlds include different instruments, strategies for prototype testing, different production capacities, different modeling techniques as well as varieties of scientific theories, concepts and principles. Working within their respective object worlds, each participant *sees* the design but *sees* it *as* differently. One object of design, different object worlds. (Bucciarelli 99)

Designing Engineers includes, as one of three case studies, the design of an x-ray inspection machine intended for use at border crossings and at airports. This was a relatively large machine intended to inspect aircraft cargo containers and even trailer truck contents. At the outset, participants in the design confront some common textual materials,- memos, client specifications, some hardware, perhaps a competitor's product. More likely the firm's own prior productions, drawings and sketches define their field of view. But how do they read and interpret this information?

That depends upon whom you ask, upon their particular responsibilities and technical interests. A mechanical engineer reads the specifications and is most interested in how they translate into the structure of the support for the sensing crystals, the track and train for motoring the cargo container through the beam of X-rays, between the X-ray source and the tower of crystal receptors. The physicist sees the proposal as a problem in determining the strength of the required beam, how it will be focused, what they will use as a source. A software development engineer focuses on how to filter, massage, and display the signals produced by the crystals to produce the best image on the computer monitor. The chief officer of the company envisions a small car being towed through the beam and the marketing advantages that will accrue. They all see, all have access to the same texts, things, drawings and at some common level, talk about these things without a problem, but when they get down to business within their respective object worlds, the "seeing as" something different is immediately apparent.

3.2 Designing: A Social Process

Now it would be advantageous if the design task could be broken down into a set of isolated object-world tasks which might be pursued independently. One can strive to reach this happy state by careful and thoughtful organization at the outset. As part of this one must define clear "interface conditions" which will govern how the pieces, often literally, will fit together. But this is a most difficult state to achieve. Because design is about making what is new, a venture into the unknown, a process where uncertainties abound, it is impossible to set out a clear and concise set of specifications at the outset upon the basis of which one might attempt to completely dis-aggregate the design task. While it remains a worthwhile venture and a necessary element of effective management, boundaries and interstices remain open and often shift. Specifications themselves, in their elaboration in process, ought to be considered subject to re-design.

One can not break up the design task into independent tasks, addressable by single disciplines alone. Likewise, there is no over-arching object world whose instruments and methods, language or tools, can reconcile the different proposals and interests of participants from different object worlds. Consequently, participants must collectively articulate their proposals and negotiate their differences. Of course, participants speak a common language. But there is no way to effect a translation, *in instrumental terms alone*, of the theories, methods and proposals of one object world into another. The challenge then, is to move beyond the different perceptions of what the product should be, what it is, different perceptions of how it works, and to harmonize all participants' interests in some way. That way is necessarily social: In this sense, design is a social process.

Given this perspective, what can we make of knowledge, of design knowing and learning and what are the implications for engineering education?

4 Design Knowledge

Knowledge, according to the dictionary is in one guise "...the fact, condition of knowing....the fact of being aware". This does not, I suggest, convey fully what we mean when we use the word. Surely we want our students to do more than be aware of Newton's Laws, that finite element programs exist, that computers contain a processor and memory chips. There is a saying "...he knows all but understands nothing". No, we want our students to have knowledge in the sense of an understanding of the fundamentals of the appropriate paradigmatic sciences *and* an ability to act, to design, with that knowledge in hand. This coupling of understanding with action is what makes attempts to define design knowledge so difficult.

Design knowledge is not information: I define *information* as inscription - i.e., as texts and graphics in hardcopy or electronic form. An engineering textbook contains information; a computer program printed out or displayed on a monitor, a drawing or sketch, a prototype of hardware - these are all information. Information is a constituent of knowledge, of knowing, but possession of it by itself alone, having it in hand, does not mean that one is knowledgeable. To possess knowledge is to be able to not only have the information in hand or recalled from memory, but to construct an interpretation appropriate for the immediate occasion. Knowledge is latent in words and texts conjoined with context. Texts and words - information - can be misinterpreted as often evidenced by our students inability to grasp the essentials of the information we convey to them.

Similarly, I distinguish *skills* from knowledge: *Skills* are learned through practice and drill. To juggle, catch a ball, ride a bicycle, solder a lead, add 2173 and 184, cook an omelet, replace a washer, do a triple by-pass surgery, drive a car, play the piano are skills. Skills are learned by doing. Skills are governed by rules. Sometimes these need not be articulated and "taught"; one can learn to juggle, ride a bicycle without rules but clearly in many cases they can be essential to developing the skill. But - and this distinguishes skills from knowledge - one does not reflect on the rules in performance. Certainly some of these fore-mentioned skills require something more than automatic performance, e.g. triple by-pass surgery. They require thought, creative thought, and judgement at times. But it is thought in action, constrained by the immediate context, the necessity of moving on with the game, the flow, the task at hand. Generally skills are in the possession of, and exercised by individuals.

Training goes with skills; one trains to develop ones skills. Reading goes with information; one reads a text, a spread sheet, a data base, even a prototype. Learning goes with knowledge. Yet, from the way some faculty speak and write about knowledge one might infer that knowledge is but information. Our traditional lecture/recitation educational format reflects this usually un-stated belief: In lecture we convey knowledge-as-information and in recitation students develop their skills applying this information in the doing exercises and projects e assign.

Knowledge, taken in this way, as information, metaphorically becomes a material substance. It is as if it were something fluid or plastic, able to be shaped and transmitted by us and received and absorbed by our students. Students might complain about the amount they are expected to learn in a semester - it's like taking a drink from a fire hydrant - but they feel short-changed if only one-half the syllabus is covered by semester's end.

In keeping with, rather possessed by, this metaphor, we think of knowledge as accumulating. We proclaim how, through our research, we produce new knowledge and add to a body of knowledge. We count the number of publications of a colleague in our tenure and promotion evaluations. Our curriculum discussions center on what courses to leave in, what to add, what to leave out.

The point is that this imagery of knowledge as substance frames the way we evaluate and understand knowledge and knowing in practice and, as such, leads us astray. Knowledge as stuff is taken as static, stored in memory, a text, a database, on the web, and available for recall and application. This imagery goes hand in hand with the instrumental, perspective that is so effective in object-world work but it ignores the contextual nature of knowledge. How could context matter if knowledge is a material existing in its own right? It discounts immediately that knowledge and knowing might depend on linguistic and social resources as well as cognitive processes; that knowing is an active, creative process and not simply a passive assimilation of established ideas; that knowledge is culturally situated and socially distributed as well as personal. (Easterby-Smith et al. 99)

There is more to knowledge than information and skills. But what is there? One way to approach this question is to raise another - how do we recognize when someone has knowledge? Generally we observe the individual in action and pass judgement. "Does so-and-so know what he/she is doing?" This is the judgement we make. Knowledge and doing are quite rightly conjoined in this test.

In 1946 Gilbert Ryle, in a famous essay, distinguished knowing *how* form knowing *that*. He made a distinction between being intelligent and (simply) possessing knowledge.

He claimed peoples' competencies, rather than their "cognitive repertoires", are the measure of intelligence. Knowing *how*, how to do, "is among other things, considering what is pertinent and disregarding what is inappropriate" (Ryle 46).

Ryle moves us away from the knowledge as stuff metaphor and toward the importance of context. For knowing what is pertinent, what not, is a matter of context. Knowing, for example, what level of detail is appropriate in a model of the behavior of the object of a design depends upon where you are in the design process. Early on, in a conceptual design phase, models can be quite crude, indeed must be so. Later more detailed models and more accurate results are usually required. Knowing *how* to model is a matter of context.

So too, judging which performance specifications are to be given priority, which are negotiable is critical. I refer again to the design of the X-ray inspection machine where, at a decisive stage in the design process, plans were being drawn up for demonstrating the systems capabilities as required by contract. One requirement stated that an explosive, hidden away in a carton or piece of luggage, was to be detectable. But what is an explosive?

Participants at a planning meeting had to fix what this substance was to be, how much of it was to be stashed away, and with whatever else in the cargo container. They debated what form an explosive might take, finally agreeing that an explosive was a packaged substance but including wires as part of the package. Without the wires, detection would be more difficult.

Now we can ask what knowledge was displayed or indicated at this meeting? On the one hand, knowing about the relative opaqueness of plastic and wire, what a bag of plastic will look like after the signals from the array of crystal detectors have been processed by the computer and massaged with contrast stretching, edge enhancement and zooming is required. Only then might one establish some sense of what an explosive must consist of if it is to be detected. On the other hand, knowing something about the intentions of the authors of the specification is important; how much flexibility is there here? Knowledge too about limits and boundaries - on budgets, on resolution, on time available, power supply - all of this is displayed in this definition of an explosive. Participants must agree, establish a common representation and presentation of the "explosive". Only then can the test item be fabricated and the testing and the design process go forward. Knowledge about explosives is context dependent, in part theoretical, in part experiential, in part social; the meaning of the word "explosive" is socially constructed.

Knowing how to employ the language of design is clearly important to full participation in the process. Here we can observe the facility with which a participant constructs, modifies a sketch in their object world work or in negotiations with others - sketches, drawings, graphical representations are ingredients of the language of design; we can watch how they translate statements about characteristics and function into the hardware and software constituting prototype. We can listen and see how ambiguity is resolved and exploited in exchange among participants. Knowing *how* to listen is part of knowing the languages of design. This too is a mix of the experiential, theoretical and the social.

Rules, of course, are important to designing: Rules as embodied in methods for making decisions, scheduling tasks, allocating resources; rules as codes and regulations, specifications and the like; and rules as design heuristics, empirical guides as well as more direct consequences of theory. But knowing the rules, being able to recite them, quote chapter and verse, does not suffice. Rules always require interpretation in application. Knowing *how* a rule is to applied, if it is to apply at all, is also context dependent, dependent upon the beliefs and norms prevailing within the team, within the firm. As such they too are socially rendered.

All of this is not to deny that skills and information are not important ingredients of good designing. (Good skills and good information). Manipulating a spread sheet, soldering a lead of a prototype, or facility with a CAD tool are prerequisite to effective design practice and data of all sorts - market estimate, suppliers specifications, material properties - is essential. Skills and information can mark the difference between an efficient and effective design process and one that runs over budget, over time. But here too, knowing *how* to exercise a skill or *how* to filter and interpret information is critical and context dependent.

The essential and all important difference between the vision of design as a social process and the vision implied in the knowledge-as-stuff talk is that the latter has no need of others. In this mode of speaking, it is the individual alone that matters, that possesses and applies skills and knowledge. There is no mention, no need of others. But design is a process that can't be so executed. Design knowledge is not stuff. That is, knowledge in designing is so dependent upon context and exchange among participants that to put it in a box, to teach as if it could be so isolated, is mistaken. It is true, we can't teach design.... if we insist on trying to teach it this way.

If we except this vision - that design is a social process; that domain specific information and skills are essential but do not suffice; that design competence is context dependent and socially articulated - certain implications follow, not just for teaching engineering design but for the whole engineering education enterprise - the content and ways we teach all courses i.e., engineering science, laboratory as well as capstone design course.

5 Implications for Engineering Education

There is a tension here, one evident in the workplace in contrasting the experience of object world work and social process. It is the tension between a naive but powerful instrumental approach which sees an engineering task as the application of the stuff of scientific knowledge and the exercise of instrumental rationality alone and an approach that sees engineering design as sensitive to cultural context. This tension is what lies behind the notion that engineering design is something different from engineering science and why some claim we can't teach design.

It is true -we can't teach design like we teach science. But to try to make design into a science is to avoid the problem, to negate the tension altogether in the classroom only to exacerbate and heighten the tension the first day on the job. Rather, if we choose to confront the tension - let me call it the tension of instrumental content and cultural or social context - allowing that it is important that our students be prepared to effectively work within the mix, then we ought not look at "teaching design" alone but critically appraise our beliefs, methods and objectives in teaching instrumental knowledge - the theories and object world practices of engineering science. How can this be improved, changed to better accord with the needs of industry, if I may be allowed the cliche?

I turn to consider the way we teach, focusing on the beliefs, methods and objectives of instrumental learning, using two simple problem to illustrate. I show how these problems might be embedded in a more open-ended context so that the tension is made explicit and occasion for learning.

5.1 An Example

If we look at the ingredients of the technical core of any engineering curriculum we see a high degree of abstraction in accord with the reductionist strategies of science. We are so accustomed to the traditional classroom context and to traditional content that it may be hard to see what we do as highly abstract or how it could possibly be otherwise. After all, these are the fundamentals, the basis of all rational engineering effort. I proceed with two examples; two exercises a student might encounter in a Mechanics course at the sophomore level.

Exercise 2.8– *Show that* the force F required to *just start* the lawn roller, of radius R and weight W, moving up over the ledge of height h is given by

$F/W = \tan \phi$

where $\cos \phi = 1 - (h/R)$

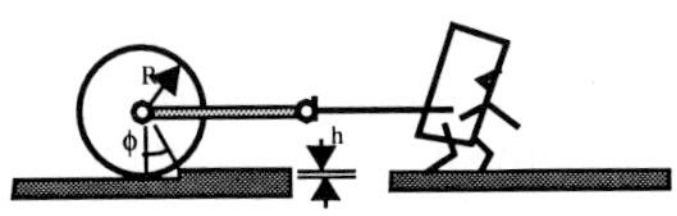

Figure 1: Traditional roller problem.

That this is an abstraction may not be so evident to one who has learned the language of engineering mechanics: The frictionless pin bearing where the handle connects to the roller; the two dimensional nature of this world; the restriction of orientation of the handle to horizontal (and consequent distortion of the he/she/it pulling); the letters designating dimensions - these are all read out of the drawing without thinking about of what this is a model. The student is no doubt familiar with lawn rollers, or at least has seen photographs or other images of this technology. But they need not worry about the relationship of this abstract representation of this particular problem to anything out there in the so called real world. They quickly come to understand that the words and figure are to be read, tested, and interpreted only in so far as they relate to finding an answer to the stated problem. The world the student must connect up with is the object world of Static Equilibrium, the abstract world of Force Equilibrium, Free Body Diagrams, Force as a Vector, and a bit of Vector Algebra and Trigonometric Functions. To one not so versed it must seem as bizarre artwork...and what about the gap in the earth?

It is as if we look with our students at the real world of rollers, structures, and machinery with a special pair of glasses - an engineering mechanics pair of glasses - which lets through only icons and images of abstract vector forces, lines of action, lever arms, angles, frictionless pins and the like. All else is filtered out. As a result, the context for problem solving is tightly drawn; who the person laboring to pull the roller over the bump is, indeed whether for large bumps he/she/it has the strength to do the job, how big the field, how much the cost, what hourly wage - all of this is irrelevant. If the student expresses curiosity about any of these realities he or she will draw blank expressions or be laughed out of the room. The setting of the technology in the world is not of concern. The real world context is ignored, rather only attended to in a most limited way, as a set-up, a cover for a problem set in engineering mechanics.

The example is meant to increase the student's problem solving skills; here is where the general rules for solving problems, found in an appendix of the textbook for the course, would seem to apply: "...read the problem; draw a free body diagram; include unknown internal and reaction forces; pertinent dimensions.....check your solution". (Gere and Timoshenko 97) Through this sort of exercise, the student acquires more and more knowledge of engineering mechanics and becomes better able to apply this knowledge (to solve more intricate problems). Here is where the fundamentals of an object world are grasped and internalized.

The classroom context of the exercise and the educational experience of the student, is generally not given much thought. Content is sacred. But what is the nature of this setting? Here is where students learn that their success in the course depends upon their individual efforts made in competition with their peers. Here is where they learn what constitutes a relevant question, what not, where they learn to see the world in a most filtered and reductionist way. Here is where they learn not to see -not to question, listen to others, create, play with ideas, argue, conjecture, estimate, nor negotiate - and not to design. Here is where the metaphor of knowledge-as-substance is inculcated, that knowledge can stand alone, independent of context.

I want now to show how the same stuff could be taught, within the same course, but in a completely different context, one that accords with the world of engineering practice in more than its object world features. The exercise I lay out below is meant to change the context of the problem and the context of learning. The objectives of the course still include the teaching of the fundamentals of engineering mechanics but that is but part of the picture. The exercise:

Exercise 2.8 (transformed) Hospital Bed – Wheel Size

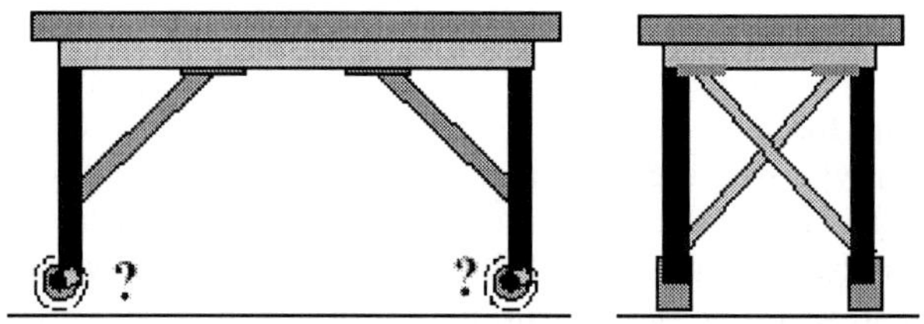

Your task is to do a first-cut analysis in support of the design of a new, light-weight, mobile hospital bed. The bed will be used to transport patients indoors on caster type wheels over relatively smooth terrain but there will be some small bumps it must traverse without discomfort to the patient. A single attendant should be able to push the bed to its destination. Develop a rationale for fixing the size of the wheels and use it to determine a range of possible diameters.

Figure 2: Transformed Roller Exercise

First note that information is lacking and intentionally so. The student is meant to grapple with the question: "What additional information do I need to respond?" And a related question: "Where might one obtain this needed information?" How you deal with this as faculty depends upon the resources you make available, or the students have available in their own right. But the name of the game is not simply to define the parameters for students and to give them the numbers but to have them develop some rational way of constructing both what measures are important and their appropriate values.

Here is an opportunity for moving from individual effort to social exchange: I have students work in groups of three or four; have them discuss among themselves what is important to know. We do some collective conjecture and estimation of the weight of patients, the likely dimensions of hospital beds, the size of bumps in a hospital, the force with which one can push. They are encouraged to go to other sources - the web, the library, even call a company that makes hospital beds. In this way they take and share responsibility for the formulation of the design task.

Once the students realize that the name of the game is not to figure out what is in the head of the teacher but rather to take responsibility for the formulation as well as carrying through of the exercise, the ambience of the classroom changes dramatically. Evidence of this change in context is found in student's questions and musings: Some will ask if how we ensure the safety or comfort of the patient with just one attendant pushing. Why can't the attendant go around to the front and pull the bed over the bump? Another will ask if costs are a constraint; how they might vary with wheel size, whether this is significant; when might this be significant?

The success of this transformation depends on the attitude and bearing of the instructor. If the exercise is to be a vehicle for design learning, as well as in statics, faculty must listen and respond creatively to these queries. No question ought to be ignored, laughed at or labeled out of bounds. Yes, it would be good if there were more than one attendant, of course; and it does seem that there would be advantages to going around to the front of the bed and pulling the patient over the bump; can you show that? But let's design for the worst case where this may be impossible or the attendant may not be as thoughtful as you are. (You might even use the occasion to talk about idiot proof design - a case of engineering arrogance or good, robust design?) Faculty must be prepared to negotiate the context, the boundaries of the problem. Even the outlandish ought to be answered: One student claimed the whole exercise was ridiculous - no one in the real world would do this sort of analysis; they would just call up a manufacturer of caster wheels and order an appropriate set of what appeared to be a reasonable diameter. Again, it is not hard to imagine circumstances, a context where the analysis would become necessary - say if the hospital was sited on the moon.

Once the task is more or less established, the stage is set for analysis: The task moves to relate the size of the wheels to the force of the attendant, the dimensions of the bed, patient weight, bump size. In short, a statics problem. Here again I have the students work together, at least to get them moving. The challenge is to both keep alive the ambience of active learning (with others) while having them grapple (as individuals) with the object world requirements of constructing an analytical model and exercising that model to produce a (range of) wheel size.

5.2 Another Example

I offer another example, again comparing a traditional rendering of a problem with "the same" exercise, opened-up and embedded in a different context. In this case, the exercise is intended to introduce students to the concepts of deformation and displacement of mechanical systems and how many systems can not be designed and/or analyzed without taking into account how a body deforms and moves, even if ever so slightly. The traditional exercise reads (Crandall, Dahl and Lardner 78)

2.1 A wood diving board is hinged at one end and supported 1.5 *m* from this end by a spring with a constant of 35 *kN/m*. How much will the spring deflect if a young man weighing 600 *N* stands at the end of the board? Will the spring deflection be altered if the board is made very rigid?

Figure 3: Traditional Diving Board Problem

The figure shows a man standing at the end of the board, 1.35 m out from the location of the spring. The spring itself is 1.5 m from the support at the other end of the board. The thickness of the board is given as 50 mm. (I do not include the figure; in terms of the figure below, the length, *L*, of the board would be 2.85 m, and *a* would be 1.5 m.)

To solve the problem, to get the single answer, the student must isolate the diving board and show all the forces acting on it - the weight of the man, the force due to the spring, the force of reaction at the right end - then apply the requirements of static equilibrium to determine the force in the spring. With the given spring constant, the student can then compute the deflection.

Design Exercise 3.1 Low-end Diving Board

You are responsible for the design of a complete line of diving boards within a firm that markets and sells worldwide. Sketch a rudimentary design of a generic board. Before you start, list some performance criteria your product must satisfy. Make a list also of those elements of the diving board, taken as a whole system, that determine its performance.

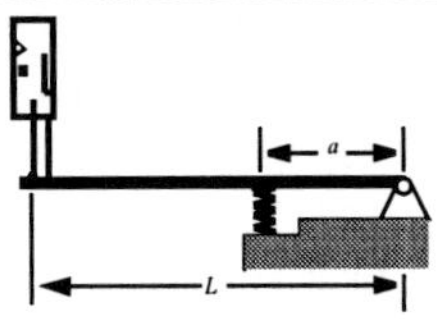

Focusing on the dynamic response of the system, explore how those elements might be sized to give your proposed design the right feel. Take into account the range of sizes and masses of people that might want to make use of the board. Can you set out some criteria that must be met if the performance is to be judged good? Construct more alternative designs that would meet your main performance criteria but would do so in different ways.

Figure 4: Transformed Diving Board Exercise

The student must also understand that it doesn't make any real difference if the board deforms i.e., bends a little. The equilibrium equations may still be written with respect to the undeformed configuration. The board may be treated as rigid. The student might be thrown off by the indication of the board thickness. Generally only information needed to pose and solve a problem is given explicitly in the problem statement. The thickness indicates that the wood board is quite hefty, reinforcing the notion that the board may be treated as rigid. Here is an instance where the student must rely upon his experience with diving boards of this type or at least realize that if the board deforms excessively as the man moves out to the end, it can no longer function as a diving board. In the traditional setting, all of this might be considered a diversion, keeping the student from getting on with the task of finding the answer. But considerations like these are not off track when the exercise is transformed into a design task.

While the figures may be identical, the way the figure is to be read is quite different. For solving the traditional, single-answer problem, the student is to put on the special glasses that enable seeing the underlying form, the analytical contents behind, within the line drawing - replacing the man (earlier editions had a woman) with a force vector, the compression in the spring with another, so too the reaction at the support. The information is all there, coded in the figures dimensions explicitly given and the numerical values in the text of the problem statement. He or she must focus and filter out all associations and reflections on diving boards as they really are (other than for the diversion concerning the rigidity of the board).

The same figure is read differently by students doing the design task. Now all questions are admissible, seen and recognized, if only but to clarify. I recall how a student straight-off questioned the existence of diving boards of this type. He had seen only the sort where the spring of the system was derived from the elasticity of the board alone; there was no spring in his experience. But others immediately chimed in: Oh yes, they had seen this type; one had such a board in her backyard in San Diego. All of this banter would be considered irreverent in the traditional classroom setting. But here, in this open context, not so: it can lead to a discussion of market segments for different types of boards, to their relative cost, all of which, while not critical content of engineering mechanics *is* critical in motivating students to take hold of the problem as their own.

Asking students to list some performance criteria prompts further questions and appropriation. Asking them to give particular attention to the "feel" of the board raises questions about the dynamic response of the system, the nature of its springiness, more specifically, what should its fundamental frequency be? I as the students to stand up. (When was the last time students stood up in the middle of a traditional recitation session?) Once standing and free of initial inhibitions, I tell them to close their eyes and imagine they are standing at the end of the board, preparing to dive. What should the frequency be? They bob up and down and report numbers in the neighborhood of 1 to 2 Hertz. This then becomes a performance specification.

In a similar way, working in teams of twos and threes they make estimates of the range of weights of users, argue whether all ages should or could be accommodated with such a simple arrangement. "Can we make the spring movable?" How you, as faculty, respond to this question will depend upon the resources students have available, specifically how much time they are allowed to work out the exercise. Contrast this with the traditional exercise -

one of three due weekly is the standard formula. Here there is considerable flexibility and again, the students have the opportunity to contribute to the formulation of the problem.

Once students have a range of weights, a limit on the deflection of the board, a number for the desired resonant frequency of the system, they can begin to figure out what the length of the board, the location of the spring, and its stiffness should be. Here I introduce the notion of equivalent stiffness, and we explore how one might model the system as a simple mass-spring system. But what about the distributed weight of the board? We talk about the possible effect of assuming the board weightless or lumping into the mass on the spring some portion of the boards weight. What about its rotary inertia?...

Clearly in an open-ended exercise, discussion, conceptual renderings, model making and adjustment could go on for the whole semester. So bounds have to be set, according to the resources available and justified by course objectives as well as estimation and approximation. Note too how I do not shy from requiring students to grapple with some fundamental notions of the dynamics of structures, though the course is defined as a course in statics alone. Again, the open-ended context allows, like the so-called real world requires, students to confront the interdisciplinary character of engineering practice.

Because of the open ended nature of the exercise, there exists more than one suitable response; and there are also alternate ways of constructing that response. This makes evaluation a more arduous job, compared to what is required to score a traditional problem set. The way the student documents their work, what assumptions they make, how they justify these, the robustness of their estimates, as well as how well they keep track of units, are consistent and complete in their application of the principles of static equilibrium - all of this must be evaluated. I have them keep a journal.

6 Conclusion

To transform the student's learning experience to better accord with design knowing, thinking, and acting, one need not add courses in design science, teamwork, negotiation skills and the like. Nor need one become too concerned if all of the material traditional listed in a course syllabus is not covered. The knowledge as stuff metaphor does not suffice. The way to better prepare students for engineering practice is to transform the context of the engineering classroom, not just in those courses explicitly labeled "design" but across the board, starting with the engineering science courses at the sophomore level.

Through the integration of open-ended, design-type exercises into the curriculum a dramatic change in the atmosphere and context of the classroom can be achieved and the fundamentals still be addressed with rigor.

A cautionary note: It is not the directives of the exercise in itself which fixes the context of the classroom and the student's experience. Consider, for example, the transformed roller exercise of Figure 2: If I do not take off my engineering mechanics eye glasses, the exercise may simply be seen as a more difficult edition of the original problem and could easily be treated as traditional. "Size the wheels" could be interpreted as "Find the size of the wheels". In this case, the instructor, in response to students' expressions of unease, would then specify the dimensions of the bed, the weight of the patient, the force with which an attendant can push and the size of the bump. Some further guidance would be needed, no doubt, in seeing students through the drawing of a free body diagram, but if they had worked on, and been shown the solution to, the original lawn roller problem, this might not take much time nor thought. A single value for the diameter of the wheel could then be deduced from this information and the requirements of static equilibrium. In short, the problem could be transformed back into a "single answer problem" with its solution, the wheel diameter, given in the back of the book. Following the problem solving rules given in the Appendix, one should be able to achieve a solution.

The point is that the information displayed in the open-ended problem statement - text and figure - might very well be read in a traditional way and the context of the classroom left unchanged. Yes, we have changed the statement of a problem but that does not necessitate a change in student nor teacher experience.

The intent, rather, is to change the rules of the game (metaphorically speaking) while using the same tokens, the same symbols and vocabulary - including figures. An analogy: Think of the several different ways in which checkers and a checker-board can be the stuff, the material objects, of different games. ("Laskers" was one such variety I remember playing as a graduate student). Or more obvious, think of the varieties of games played with the same set of fifty-two cards of four suits. Some games are simple, child's play; others sophisticated and complex, requiring teamwork.

What games do we want to play with our students? Those that better accord with the world of engineering practice ought be our aim. That engineering design requires more that the application of science-based theories and instrumental reasoning is recognized by faculty and practitioners alike. But how to teach the "more" requires a new kind of game, one that recognizes that "knowledge" in engineering practice is very much context dependent and socially construed.

References

Bucciarelli L, (1994), *Designing engineers,* MIT Press, Cambridge, MA.

Bucciarelli L L, (1999), Design Delta Design: Seeing/Seeing as. *4th Inernational Design Thinking Research Symposium on design representation*, Institute of Technology, Cambridge, MA, April, 1999.

Bucciarelli L, and Schön D, (1987), Generic Design Process in Architecture and Engineering: A Dialogue Concerning at least Two Design Worlds, *The Study of the Design Process: A Workshop*, February, 8 - 10, 1987, Oakland, CA.

Crandall S, Deal N, and Lardner T, (1978), *An introduction to the mechanics of solids*, 2nd ed., McGraw-Hill.

Dixon J, (1966), *Design engineering*, McGraw-Hill, New York, NY.

Easterby-Smith, Araujo, and Burgoyne, Eds. (1999), *Organizational learning and the learning organization: Developments in theory and practice*, Sage Publications, London.

Geertz C, (1973), *The Interpretations of Cultures.* Basic Books, New York, NY.

Gere J, and Timoshenko S, (1997), *Mechanics of materials*, 4th ed., Boston, PWS Publishing Co.

Hubka V, and Eder W E, (1988), *Theory of technical systems*, Springer-Verlag, Berlin.

Lave J, and Wenger E, (1991). *Situated learning. Legitimate peripheral participation,* Cambridge University Press, Cambridge.

Nardi B A, Ed. (1999), *Context and consciousness: Activity theory and human-computer interaction*, MIT Press, Cambridge, MA.

Orr J, (1990), Sharing knowledge, celebrating identity: community memory in a service culture, Middleton and Edwards Eds. *Collective Remembering,* Sage, London, 169-89.

Ryle G, (1946), Knowing how and knowing that, *Proceedings of the aristotelian society*, vol XLVI.

Statnikov R B, (1999), *Multicriteria design: Optimization and identification*, Kluwer. Boston.

Index

E

F

G

I

K

L

M

N

O

P

Q

R

S

T

U

V

W